The Triumph of the Fungi

The Triumph of the Fungi

A Rotten History

NICHOLAS P. MONEY

2007

OXFORD

UNIVERSITY PRESS

Oxford University Press, Inc., publishes works that further
Oxford University's objective of excellence
in research, scholarship, and education.

Oxford New York
Auckland Cape Town Dar es Salaam Hong Kong Karachi
Kuala Lumpur Madrid Melbourne Mexico City Nairobi
New Delhi Shanghai Taipei Toronto

With offices in
Argentina Austria Brazil Chile Czech Republic France Greece
Guatemala Hungary Italy Japan Poland Portugal Singapore
South Korea Switzerland Thailand Turkey Ukraine Vietnam

Published by Oxford University Press, Inc.
198 Madison Avenue, New York, New York 10016

www.oup.com

Library of Congress Cataloging-in-Publication Data
Money, Nicholas P.
The triumph of the fungi : a rotten history /
Nicholas P. Money.
p. cm.
Includes index.
ISBN-13 978-0-19-518971-1

1. Fungal diseases of plants—History.
2. Fungi—History.   I. Title.
SB733 .M59 2006
632'.4—dc22   2005037223

For Adam, my stepson

Preface

This book is concerned with the most devastating fungal diseases in history. These are the plagues of trees and crop plants, caused by invisible spores that have reshaped entire landscapes and decimated human populations. Everyone is aware of the Irish potato famine, but while many other fungal diseases are less familiar, they have had similarly disastrous consequences. *The Triumph of the Fungi* focuses on the fascinating biology of the well-known and lesser-known diseases. It also tells the stories of the scientists involved in their study and of the people directly affected by the loss of forest trees including the chestnut, and cash crops such as coffee and cacao. Although a book about fungal epidemics isn't tailor-made for an intoxicating and uplifting read, the chronicle of the mycologists and plant pathologists engaged in combatting these diseases is one of human optimism (often encouraged by desperate eccentricity). In a surprisingly brief time, human knowledge of the fungi that infect plants has evolved from Biblical superstition to the recognition of the true nature of plant disease and, more recently, to a sense of awe for the sophistication of these organisms. The crucial issue of human culpability in these fungal epidemics is addressed in the book's closing chapter.

A note about the title of the book seems appropriate. In the second year of World War II, the engineer, novelist, and plant pathologist Ernest C. Large published a marvelous book, *The Advance of the Fungi* (New York: Henry Holt and Company, 1940). Large introduced scientists to the study of plant diseases with a refreshing mixture of technical rigor peppered with humorous asides. *The Advance of the Fungi* served as the introduction to fungal biology for many of the plant pathologists that staffed university departments and government-funded laboratories throughout the second half of the twentieth century. A second book, *Mushrooms and Toadstools* (London: Collins, 1953), published in 1953 by John Ramsbottom, served as a similarly good-humored and inspiring source for mycologists. Rams-bottom's book served as a model (albeit unconsciously) for my first book, *Mr. Bloomfield's Orchard* (New York: Oxford University Press, 2002). The focus of both works, separated by a half-century of discoveries, was on top-ics such as fungal growth and mushroom function. In similar fashion, *The Triumph of the Fungi* updates Large's classic by offering a personal view of the continuing advance of the fungi in the last 65 years and by revisiting the history of the scientific study of plant disease.

I think that Large would approve of the new title. Since 1940, fungi have continued their advance, attacking every crop plant that we cultivate, and exploiting new hosts wherever spores are introduced. Through their con-tinued advance, the fungi have proven unstoppable. Fungi are the most important cause of plant disease and cause billions of dollars of crop losses every year. Despite fantastically effective fungicides, the continual develop-ment of resistant varieties of crops, and the implementation of techniques of genetic modification, blights, rusts, and rots abound. After more than a century of concerted scientific effort, epidemics like potato blight, chestnut blight, and Dutch elm disease remain incurable. The best we can do is to continue the expensive fight to limit the negative consequences of fungal activity throughout the biosphere. On a more positive note, biologists have been successful in documenting the essential nature of our varied interac-tions with fungi. It is clear that we would find the planet uninhabitable without fungi.

The presentation of the epidemic diseases in *The Triumph of the Fungi* does not follow their historical appearance nor the history of their recogni-tion by humans. Instead, the book opens with the story of chestnut blight, the fungal disease that reshaped the forests of the eastern United States in the twentieth century (chapter 1). It is difficult for us to appreciate the overwhelming impact of this disease for the simple reason that few of us were born early enough to have seen a giant American chestnut. (Inciden-tally, 2006 is the 100th anniversary of the first description of the blight fun-gus.) Chapter 2 describes an equally destructive fungus that annihilated elm trees a few years after the appearance of chestnut blight. More than any other fungal epidemic, Dutch elm disease has changed the appearance of villages, towns, and cities in Europe and North America. Chapters 3, 4, and 5 address fungal epidemics of three tropical commodity crops: coffee, cacao, and rubber. These diseases are united by the spread of plantation agriculture by European colonists in the nineteenth century, and they illustrate the extreme vulnerability of monoculture agriculture to fungal attack. The origins of the scientific study of plant diseases are addressed in chapters 6 and 7, beginning with the attempted placation of the Roman mildew god, to seventeenth-century experiments on plant diseases and the eventual development of the branch of science called plant pathology. The diseases that were responsible for the birth of plant pathology were the smuts and rusts of cereal crops (chapter 6) and the potato blight pathogen that caused the Irish famine (chapter 7). The final chapter (chapter 8) explores the future of the ongoing competition between humans and fungi for control of the biosphere. The fossil record shows that fungi have lived in intimate associations with plants for the last 400 million years. Although many of the earliest fungi engaged in mutually supportive relationships with land plants, others were probably attacking plants in the Silurian mud in much the same way pathogens do today. But although there may be nothing truly novel about emerging epidemic diseases such as sudden oak death, this perspective does little to alleviate concern about the future health of forests or the effects of fungi on agriculture. I hope that you'll enjoy my take on the stories in this book as much as I have relished delving into this rich archive of microbiology.

Acknowledgments

The only formal course in plant pathology that I took as a college student was taught by Julie Flood at the University of Bristol. This was, I recall, a fantastic class and Julie bears no responsibility for any deficits in my understanding of plant pathology. Many of my other tutors—including Marshall Ward, Anton de Bary, and Miles Berkeley—are long dead, leaving me to comb through their dusty writings in Cincinnati's incomparable Lloyd Library. This book would not have been possible without the Lloyd collection, and my sincere thanks go to Maggie Heran and the staff at the library for leading me through the last four centuries of plant pathology literature. A rare gap in the Lloyd's collection was filled by the American Museum of Natural History in New York City, whose librarian, Ingrid Lennon, was helpful in locating a particularly obscure journal article. I am grateful to my colleagues Holger Deising (Martin-Luther-University Halle-Wittenberg), Sophien Kamoun (The Ohio State University), and David Rizzo (University of California) who furnished illustrations. Mike Vincent, curator of the Willard Sherman Turrell Herbarium at Miami University, answered many questions and found the lovely drawing of the elm bark beetle used in chapter 2. Carolyn Keiffer, also at Miami University, shared her expertise as a chestnut blight researcher and accompanied me on a harrowing flight in a small plane to Wisconsin so that I could see surviving trees. Finally, I thank my editors, Diana Davis and Peter Prescott, and an anonymous reviewer who was responsible for the removal of some of my more irresponsible asides.

Contents

The Triumph of the Fungi

CHAPTER 1

Landscape Architect

The tiniest wound, just a nick in the bark; spores drift in from the moist forest air; later, the stem swells and ruptures; the sap-flow system is crippled, and leaves are shed. Bereft of its plumbing and solar panels, the crown of the gigantic, alabaster ghost sways brittle in the breeze. When this story was repeated 3 billion times, a tree that had dominated the eastern woods of North America for more than 50,000 years was exterminated. This is what happened to the American chestnut, *Castanea dentata* (fig. 1.1), when it met the landscape architect called chestnut blight.

The beginning of the end for the American chestnut is usually tracked to a report by Hermann W. Merkel, a forester who discovered dying trees at the Bronx Zoo in the summer of 1904. Foliage was wilting on diseased branches whose bark revealed telltale bands of infected tissue. Merkel said that the disease was limited to "a few scattered cases" in 1904, but by the next summer the blight had spread. Chestnuts of all sizes showed the same symptoms: the disease had escaped from the zoo. Merkel estimated that 98 percent of Bronx chestnuts were infected. The chestnuts grew in parks throughout the borough and lined avenues just as they graced and shaded cities across the country. Merkel requested $2,000 to engage a crew to prune and burn diseased branches from trees in the zoo. This seemed like a good defense: gardeners often succeed in keeping fruit trees productive by trimming their scabbed limbs. He also purchased a power-spraying machine with a 150-gallon tank for $175 from the Niagara Gas Spraying Company and doused trees with a "potato strength" (strong) solution of copper sulfate and lime (fig. 1.2). This compound, called the "Bordeaux mixture," was originally used to control grape mildew and had become the front line fungicide of proven effectiveness against diseases of fruit trees

Fig. 1.1 Leaves and fruit of American chestnut, *Castanea dentata*. From F. A. Michaux, *The North American Sylva; or, A Description of the Forest Trees of the United States, Canada, and Nova Scotia. Considered Particularly With Respect to their Use in the Arts and their Introduction into Commerce. To Which is Added A Description of the European Forest Trees.* English translation, vol. 3 (Philadelphia: D. Rice and A. N. Hart, 1857).

and a variety of crops. The task was monumental. Three men operating the sprayer could treat no more than four mature trees in a day, exhausting an entire tank of fungicide, and there were hundreds of chestnuts on the grounds of the zoo. Merkel's team managed to prune the diseased branches from more than 400 specimens, carving some of the trees down to nothing but a bare trunk. But despite these efforts, Merkel was realistic about the likely outcome: "Just how far we have checked the progress of the disease is a matter of conjecture until the growing season reveals the facts. Considering, however, the ease with which the spores may be transferred by the action of the wind or by squirrels, birds, and insects . . . it is much to be feared that no permanent results will be achieved except by concerted action on the part of all of the Park authorities in this Borough."[1]

Merkel was amazed by the virulence of the blight. Branches of a tree that showed wilting symptoms on one day were often dead within three weeks. Trees whitened from their crowns to the ground in a single season. Massive

Fig. 1.2 Spraying an infected American chestnut tree in the New York Zoological Garden in 1905. From H. W. Merkel, *Annual Report of the New York Zoological Society* 10, 97–103 (1905).

specimens were debilitated as if struck by lightning. The disease was new to science.

Enter Dr. William Alfonso Murrill, a mycologist who had just been hired as an assistant curator at the New York Botanical Garden that straddles the Bronx River immediately north of the zoo. Merkel had sent samples of the diseased branches to the United States Department of Agriculture in Washington, D.C., appealing for information. Discouraged by the lack of interest among the government plant pathologists, and witnessing the number of infected trees skyrocket, Merkel visited Murrill in the summer of 1905. The 35-year-old mycologist immediately went to the zoo and began collecting specimens of affected tissues. The appearance of the blight in his neighborhood, coupled with the apathy of the USDA scientists, afforded Murrill a fantastic professional opportunity. Forty years later, in his autobiography, he wrote that the USDA scientists had never forgiven "the 'young upstart' for beating them to it."[2] In 1906 he published a pair of classics in the literature of plant pathology, describing the fungus that was responsible for this new disease, later called chestnut blight.[3] Murrill also reported the sad news that the disease had been identified in New Jersey, Maryland, the District of Columbia (visible from the office windows of the USDA scientists), and Virginia. Others had reported blight in Alabama and Georgia. Because all of these outbreaks occurred less than a year after its appearance at the zoo, the zoo couldn't have acted as the original source—no fungal disease could spread that fast. The origins of the developing catastrophe were a mystery.

In his scientific papers, Murrill drew the spore-producing structures of the fungus and reasoned that trees were probably infected through wounds in the bark, or through the natural breathing holes called lenticels. He also isolated a fungus from infected chestnuts and grew it on agar in glass tubes. To test whether he had found the cause of the blight, he then used the cultures to deliberately infect saplings in a greenhouse. Sure enough, the trees showed the familiar disease symptoms and began shedding a new crop of infectious spores four to six weeks after infection. The scope of his initial work was impressive. He recognized that spraying Bordeaux mixture was useless against a microorganism hidden inside the bark, and therefore any effective disease treatment would need to be preventive. Finally, he named the fungus: *Diaporthe parasitica sp. nov.* (*sp. nov.* for *species nova* or new species), which was later changed to *Endothia parasitica* and is now known as *Cryphonectria parastica* (fig. 1.3). (I use the modern name for the rest of this chapter.)

Cryphonectria parasitica belongs to a group of fungal pathogens that cause a variety of diseases including cankers of soybeans and peach trees, stem-end rot of citrus fruits, bitter rot of grape, and dogwood anthracnose. Related fungi produce toxins inside plant tissues that poison farm animals. Goats develop a disease called lupinosis after consuming toxin-containing moldy lupins, which is characterized by symptoms of heavy breathing and depression (how is this manifested in a goat?), followed by the animal's lapse into a state described as "comatose with snoring," all according to the website www.goatwisdom.com.

By 1908 the disease had caused millions of dollars of damage to trees in New York City, leading Murrill to conclude that there were no effective treatments.[4] Even the kind of careful pruning tried by his colleague Merkel

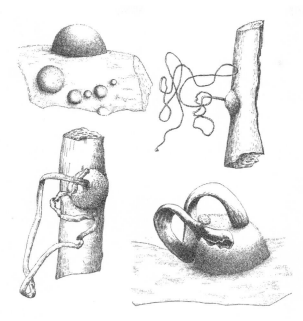

Fig. 1.3 Slimy tendrils of conidia (asexual spores) extruding from pustules of the chestnut blight fungus, *Cryphonectria parasitica*. From W. A. Murrill, *Journal of the New York Botanical Garden* 7, 143–153 (1906).

was useless. Reports of diseased trees arrived from all along the eastern seaboard, and another chestnut species called the chinquapin (*Castanea pumila*) was also falling prey to the blight. Merkel suggested that diseased trees be cut as soon as possible so that their lumber could be salvaged before further decay.[5]

American chestnuts once grew to an enormous size. In the southern part of their range, the trees reached a height of 24–37 meters (80–120 feet) while swelling to a diameter of 1.5 meters. They had been called the "redwoods of the East." Exceptional specimens of the tree measured in the nineteenth century exceeded a circumference of 10 meters.[6] Though photographs show people dwarfed by the magnificent trees, chestnuts never came close to competing with Californian coastal redwoods that can tower more than 110 meters. But nothing grew taller than *Castanea dentata* from Maine to Alabama in a 500-kilometer-wide diagonal of deciduous jungle spined by the Appalachians. The impact of the near extinction of the tree can be assessed, of course, from many perspectives. Chestnuts served as a staple construction material for American pioneers. The timber is very rot resistant (which is ironic given the species' vulnerability to fungal destruction), so cabins were assembled from trimmed logs and roofed with chestnut shingles. Thousands of miles of fencing were erected from chestnuts, the wood was used for railroad ties, and, closer to the end of the chestnut's history, it was used for telegraph and telephone poles. Chestnut was also used for furniture and "buryin' boxes."[7] In the old-growth forests, chestnuts may have accounted for one in four of the big trees. After the near-total deforestation of many of the eastern states by the close of the nineteenth century, chestnuts may have become even more plentiful. This is because the tree grows very swiftly, and it outcompeted other hardwoods in the areas where forests were allowed to regenerate.[8] There were billions of chestnuts before the European invasion, and there were billions, albeit younger ones, after the wave of humanity had swept from coast to coast.

Besides the value of the wood, chestnut bark was essential for tanning leather. Like the production of toilet paper, tanning is one of those easily-ignored but crucial human activities. Animal skin must be tanned before it can be turned into shoes, belts, jackets, underwear for British aristocrats, and upholstery. Once the hide has been flayed from an animal, it will decay quite quickly if it isn't dried, and when dried will be of little use as a motorcycle jacket unless the biker wants to look like a sandwich board. But if the skin is treated with tannins in tree bark to modify its collagen fibers, it becomes pliable and the fabric can be cut and sewn into something usable. Tannins are complex molecules found in plant tissues that, among other activities, bind tightly to proteins. The color of tea is derived from the tannins extracted by steeping the leaves in hot water. There are lots of tannins in chestnut bark, which is why the tree was harvested to meet the demands of the leather industry. Hemlock trees were the preferred source of tannins, but the rapid decimation of groves of this conifer in the nineteenth century led to the increasing use of chestnut. Thanks to the blight, tanneries were gifted hundreds of thousands of tonnes of tannins in the 1930s, but that was the end of tanning with chestnut in the United States.[9]

As everyone knows, chestnut seeds are roasted in great numbers on open fires as Santa Claus performs his annual miracle. Most chestnuts sold by today's street vendors come from Europe, though nuts from hybrid trees are also marketed by specialty growers in the United States, including Demarvelous Farms in Delaware. The disappearance of nuts in the forests had a tremendous impact on wildlife. Chestnuts had offered a large crop every year that fattened wild turkeys, deer, squirrels, black bears, and other animals. Acorns replaced the chestnuts after the blight, but fruiting in oaks is subject to annual variation, which leaves animals up the proverbial creek on a regular basis. Overall, one can say with some justification that the chestnut blight wasn't a good thing for North America. That was one of the conclusions reached by delegates at a conference on the disease called by the Governor of Pennsylvania in 1912.[10]

In 1911, the Pennsylvania legislature had appropriated $275,000 ($5.2 million adjusted for a century of inflation) for analysis of the extent of the blight and to determine the most effective control method. The money was administered by the Chestnut Tree Blight Commission, whose office was set up, with some irony, in a building at the intersection of Broad and Chestnut streets in Philadelphia. Treatment of the blight was a serious political issue. Before the arrival of the fungus, the value of the chestnut in Pennsylvania was estimated at $70 million, which is equivalent to $1.3 billion in today's money.[11] Having dealt with New Jersey, the chestnut fungus had already infected trees in eastern counties of Pennsylvania and was racing westward across the Appalachians.[12] The epidemic hadn't touched the woodlands of Ohio, West Virginia, North Carolina, and all points west, but anyone who had seen the forests farther east was worried.

The conference marked a triumph of emotion over science, beginning with a call to arms by John K. Tener, governor of Pennsylvania: "The time to act is now, and not after the scientific world has more fully worked out the history and pathology of the disease."[13] USDA scientists spoke about the nature of the pathogen, how it reproduced and how its spores were spread; others talked about the origins of the disease and how it might be controlled. So far so good. But a paper offered by Harvard botanist William Farlow allowed others to ignore the gloomy, but rational outlook of those who best understood the disease. Farlow disputed Murrill's description of the new species, saying that the fungus had been recognized in Europe for 50 years where it grew on a variety of trees and caused no damage. Murrill attended the conference, but his opinion of Farlow's work was not recorded by the stenographer. (With a century of hindsight it is clear that Murrill's work and conclusions about the blight were flawless.) Other respected scientists made significant blunders. The plant pathologist George Clinton said that the fungus was a native American species whose proliferation was caused by weather conditions. He also thought that the blight would decline naturally if the diseased trees were left alone.

The majority of the delegates thought that the disease could be controlled. Merkel, the Bronx forester who discovered the disease, was a celebrity at the conference and commented that his "fondness for trees in general is the only reason that brought me here; but that I should be pushed into the limelight thus—a modest violet like I am—was not my intention." He supported the commission's strategy of cutting infected and healthy trees in a wide belt to create a disease-free barrier that would halt the fungus: "Sometime [sic] ago I wrote that only when we considered a tree that is dangerously infected with an insect or a fungous pest as dangerous as a person infected with smallpox or as a rabid dog, will we get rid in our forests of insect and fungous pests."[14]

Murrill disagreed. Having been first to identify the pathogen and understand how it was disseminated, he recognized the probable futility of creating a chestnut-free barrier to starve the fungus in the Appalachians. Fred Stewart from the New York Agricultural Station concurred and stood out as one of the strongest voices of reason at the conference: "My views are so much at variance with what I conceive to be the sentiment of this Conference that I hesitated somewhat to present them. I feel like one throwing water on a fire which his friends are diligently striving to kindle."[15]

Stewart contended that the Blight Commission was "rushing into this enormously expensive campaign against the chestnut bark disease without considering as carefully as we should the chances of success." A field test conducted in Washington, D.C., in which a wide belt of diseased trees had been removed, suggested that movement of the fungus might at least be slowed, but Stewart argued that without a controlled experiment there was slight reason for optimism. Stewart's talk was followed by a profound silence. Pennsylvania's Deputy Commissioner of Forestry, I. C. Williams, went further, saying that the state had "thrown two hundred and seventy-five thousand dollars into a rathole." But the optimists held the day. Stewart and other critics were ridiculed: "It has been suggested that we should do nothing to counteract the ravages of the chestnut tree disease because we are not fully informed as to how to proceed. That is un-American. It is not the spirit of the Keystone State, nor the Empire State, nor the New England States," and on and on waxed a former Commissioner of Agriculture for New York State.[16]

The Blight Commission engaged a force of up to 200 men who would scout western Pennsylvania for diseased trees. They were given the authority to destroy trees on private property even if the owner refused to cooperate. Had the policy of eradication continued, this would have led to some interesting legal challenges to the commission's work on the basis of the constitutional rights of landowners. At the end of the year, the commission reported that there was still hope and discussed the value of pumping formaldehyde into diseased trees.[17] The commission was also enthusiastic about a contraption called the Fitzhenny-Guptill machine that looked like a brewery on wheels. This appliance was drawn into the woods and used to drench trees with the fungicidal Bordeaux mixture. People continued to believe that a spray would magically cure the blight. The public's gullibility and concern about the disease were exploited by bogus tree surgeons who injected colored solutions into trees and spread secret remedies on the soil,[18] which recalls the contemporary fraud associated with the remediation of indoor mold problems.

In July 1913, Governor Tener vetoed a bill appropriating money for the continuing work of the Blight Commission.[19] The enormity of the cost, coupled with growing recognition that the blight would spread with or without the program of eradicating diseased trees, led Pennsylvania to surrender to *Cryphonectria*. The victorious microbe slipped across the Ohio border and within a few decades had colonized the entire natural range of the American chestnut. John Holmes, State Forester for North Carolina, mourned the loss of chestnuts throughout the Appalachians, referring to the blight as "not only a State but a National calamity."[20]

ॐ

Within two years of the appearance of chestnut blight in the Bronx, William Murrill had succeeded in unraveling the major steps in the pathogen's life cycle. It was always possible that a clear picture of the way the fungus reproduced and spread would lead to an effective approach to controlling the disease. The fact that chestnut blight has, even after a century, outwitted everyone doesn't diminish Murrill's accomplishments. Here's how *Cryphonectria parastica* works. Once the spores of the fungus enter a fissure in the bark, they germinate and form the characteristic branching colonies of filamentous hyphae called mycelia. In chestnut blight, the mycelium develops as a dense white web that fans out through the plant's bark and the tissues immediately beneath.

A brief lesson in plant anatomy will help explain how the fungus kills the tree. A tree is composed of a series of concentric cylinders (fig. 1.4).

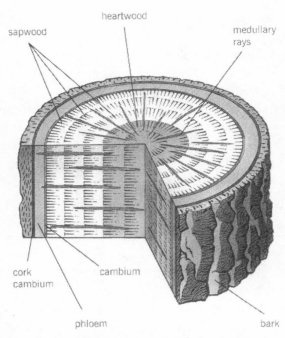

Fig. 1.4   Diagram showing the arrangement of tissues in a tree trunk. From R. A. Ennos, *Trees* (Washington, DC: Smithsonian Institution Press, 2001), with permission.

Wood occupies the interior of the tree and accounts for almost all of the trunk's bulk. The central part of this tissue is called the heartwood. The heartwood doesn't conduct water any longer; it used to, and the annual rings tell us when, but it was replaced by younger and younger wood that develops on the outside. The younger wood is called sapwood. Water and dissolved minerals are pulled up through pipes in the sapwood, called xylem vessels, that run all the way from the roots to the leaves. The sapwood is surrounded by a thin cylinder of critical cells called the cambium. This is the tissue that produces a new ring of wood every year from its inner surface. The cambium also generates the phloem, which is the second type of conductive tissue that forms a cylinder on the outside of the cambium. The phloem transports dissolved sugars in the opposite direction from the xylem. Sugars are produced in the leaves by photosynthesis, and must be moved downward so that all of the tissues in the rest of the plant are fed. Finally, the bark forms the outermost cylinder of tissue pressed onto the phloem. The usual flow of water up and sugars down is interrupted in the spring, when sugars stored in the root system during the winter are mobilized to feed the new crop of leaves. Then, the xylem that normally carries the water conveys syrupy sap, which explains why maple syrup drips from the end of a metal tap hammered into the sapwood. Chestnut trees work in the same way as sugar maples, though nobody, not even the pioneers, used chestnut sap for dousing waffles. It is the intricate vascular design of a tree that allows these static organisms to attain such massive dimensions, drawing fluid many meters into the air without a single muscle. But this superbly efficient mechanism has a fatal flaw. Disruption of the flow of water and sugars in the sapwood and phloem is lethal; there is no need to attack the bulk of the plant to ensure its demise.

The blight fungus inflicts an injury upon the chestnut from which no tree can survive. It kills the cambium. By destroying the actively dividing cells of this generative tissue, the fungus halts the production of new sapwood and new phloem. The result of this so-called girdling process is the immediate wilting of the leaves, followed by their shriveling, browning, and falling. Ringing a stem in this fashion is a very effective way to kill a tree. American pioneers defoliated trees by girdling them with an axe, which allowed them to grow crops—as soon as the shadowing leaves were lost—without the labor of felling the forest. The way that girdling works was figured out by the seventeenth-century Italian scientist Marcello Malpighi. Better known for his work on human anatomy and the discovery that the arterial and venous blood flow is linked via capillaries (he was the first to describe capillaries), Malpighi spent a decade wrestling with plants before publishing a landmark study titled *Anatomes Plantarum* in the 1670s. Malpighi girdled saplings and observed that the bark swelled above the cut and died below the cut, hypothesizing that materials manufactured in the leaves (sugars) were transported downward in the bark. He didn't know about phloem, but since this tissue is associated with the bark as I've explained, he interpreted things correctly. The leaves above the cut didn't shrivel for a while because Malpighi's shallow girdles didn't kill the cambium nor the underlying xylem. *Cryphonectria* was more destructive than Signor Malpighi.

The concept of a circulatory system in trees is unfamiliar to many people and reflects widespread uncertainty about the way that plants work (which is silly when we consider human reliance on them). As an educator, I find it helpful to get my students to empathize with the subject at hand, particularly when talking about plants and fungi. An arm serves as a convenient experimental model for a plant stem, so I propose that you begin the following thought experiment by tying a string around one of your arms just above the elbow. Notice how the veins begin to bulge in your forearm; run your fingers along the vessels and feel the valves that prevent blood backflow (it needs to return to the heart). Now, if you were to take a handful of sleeping pills and nod off for a few hours, you might make it until morning without loosening the knot. Upon waking, you would see that the hand attached to the end of the tied arm looks really nasty. If you could keep the tourniquet on tight for long enough, the limb would become gangrenous and eventually fall off, offering a perfect illustration of the way that chestnut blight girdles a tree limb. It kills healthy tissues, just as you might have done, by damaging a finely tuned circulatory system.

The chestnut blight fungus is an epicure that limits itself to the cambium. Its colonies fan out in flat plates, turning the cambium into a gelatinous mess and plugging the sapwood,[21] strangling their fragile prey. As I mentioned earlier in relation to the pioneers' use of chestnut logs, the wood shows remarkable resistance to rot. By some quirk of the chestnut's chemical makeup, mushroom- and bracket-forming fungi that spend their lives decomposing the trunks of other trees cannot dissolve chestnut logs, and so the remains of the ancient trees can still be found if you know what to look for. A pair of massive specimens lie in the woods close to my home in Ohio. I counted 80 rings where one of the fallen trunks was cut to make way for a path, and estimate that the other tree was more than a century old when it fell. They must have fallen decades ago, and probably died far earlier than that, but you wouldn't guess this by looking at the logs. A beech tree that collapses in these Ohio woods dissolves in a few years, but the chestnuts lie as if they fell yesterday. Lengths of the deeply fissured bark remain glued to the logs. Chestnut wood is ring-porous, meaning that big, water-conducting xylem vessels are produced every spring. A century after they conveyed their last streams of water skyward toward canopies of sawtoothed leaves, the vessels in the skeletons are clear as whistles, forming ring after pristine ring.

Branches infected with the blight have a brightness to them that shines against the surrounding olive-green bark. As the cambium dies, the surface of the branch sags in some areas and swells and cracks open in others. The infected areas are called cankers.[22] Within a short time interval, perhaps no more than three weeks, the fungus begins producing spores called conidia from pustules that appear as pinhead-sized orange dots: "the color of raw sienna, turning a dark umber with age," according to Merkel in 1905.[23] Each conidium contains a single nucleus, whose chromosomes are identical to those throughout the mycelium spreading under the bark. This is a form of clonal or asexual reproduction. The pustules contain chambers called pycnidia that ooze an orange-yellow paste of spores and mucilage that dries into curly tendrils (fig. 1.3). Each of these chambers can yield an astonishing 100 million infectious spores. How are the spores spread? Good question. Rain splash carries clumps of the microscopic bastards (empathizing with the tree for a moment) for short distances, but some other agent was needed to drive chestnut blight across the United States at an average speed of 3 kilometers per month for the next half century.[24]

Enter the woodpecker. Murrill had mentioned birds as potential agents for dispersing the spores in his first paper on the blight, but this idea wasn't subjected to a critical test for a few years until plant pathologists decided to shoot a variety of birds as they left blight-infected trees to find out if spores were clinging to their feet and feathers.[25] In short order, the pathologists bagged 36 woodpeckers, flickers, sapsuckers, tree creepers, juncos, and nuthatches in Pennsylvania.[26] By washing the dead birds and culturing fungi from the wash water, the pathologists found that many of the birds were carrying blight spores before they were peppered with lead; each of a pair of downy woodpeckers, for example, carried more than half a million spores. The pathologists' paper could have concluded that the only sensible strategy was to kill every woodland bird in North America. Actually, the report didn't conclude this at all, but petered out with details of lab methods for growing fungi. Fortunately for birds, bird lovers, and the fungus, but unfortunately for chestnuts, the spores were also carried from tree to tree by insects and other animals. Furthermore, long-range spread of the disease was greatly enhanced by the fact that the fungus produced a different kind of spore designed for dispersal by wind. Game, set, and match to the fungus.

The airborne spore of *Cryphonectria* is a sexual spore that is generated after the microorganism has been fertilized by a compatible mate. The sexual union is consummated when a conidium from one strain of the fungus (designated the male) fuses with the cells of another strain (the female) already established in the bark. At least in the laboratory, a pair of conidia from different strains can fertilize the colony of a third distinct strain.[27] Whether a female fungus mates with one or two males at once, the infected areas of the tree in which "she" lives begin producing a new type of fruiting body. Let's take a look with the microscope. If a thin slice of infected bark is mounted on a glass slide, a series of prominent ovals are visible beneath the surface. These are the bottoms of fruiting bodies called perithecia, and each perithecium extends a slender black chimney to the outside that will serve as an escape route for blight spores (fig. 1.5). Masses of the second kind of spore are generated inside cylinders called asci attached to the base of the fruiting body. As the asci mature, they detach and squeeze up the chimneys until they poke through the opening. Each ascus then operates as a pressurized gun that spurts a clutch of eight spores into the air above the canker, and the clouds of microscopic particles are swept away in the wind.

In the science-fiction movie *Alien*, the astronauts served as incubators for baby monsters. Remember the scene where the elaborately toothed serpentlike parasite bursts from the chest of a victim? The fungus behaves in a comparable fashion in chestnut trees, minus the attentions of a sweaty, t-shirted Sigourney Weaver. The similarity in life cycles isn't surprising because any pathogenic organism must escape from the tissues of its host if it's going to find another victim. Without an unending pattern of serial killing, a pathogen's destructiveness is limited to an act of murder-suicide—the evolutionary merits of which are no better, of course, than enjoying a hearty meal and leaping into a bonfire. (That's this weekend dealt with.)

It was the combination of masses of relatively short-range, raindrop-insect-bird-dispersed conidia and long-range ascospores that contributed to the fast and intensive decimation of the American chestnut over such a vast territory. Cutting down trees in one county might reduce the spread of the disease in the local area, but those aerially dispersed spores could float over the chestnut-free zone, traveling mile after mile with ease, to act as homesteaders for the blight in the next stand of intact forest.

Aside from the feverish investigation of the blight's life cycle and ways of treating the disease, there was a lot of interest in figuring out where this virulent tree disease had come from. Plant pathologists suspected that the fungus had arrived from another country. The only alternative explanations were that the disease had been around for a long time, but had been overlooked, or that a new virulent strain of a fungus had evolved. Given the

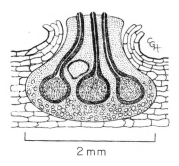

2 mm

Fig. 1.5   Perithecia of the chestnut blight fungus, *Cryphonectria parasitica*, whose necks burst through the bark. The perithecia contain multiple asci like the one shown on the right. Each ascus expels eight ascospores. From R. T. Hanlin, *Illustrated Genera of Ascomycetes* (St. Paul, MN: APS Press, 1990), with permission.

devastating effects of the disease, it seemed unlikely that it had been missed by foresters. The evolutionary explanation was also less likely than the importation theory, because there was no precedent for this kind of swift emergence of a novel tree pathogen. A related fungus called *Endothia radicalis* was known from southern Europe, where it was common on the bark of chestnut trees. But this organism was benign. It thrived on the branches but never killed the trees. The European microbe also had a distinctive appearance, which was why the cause of the blight in America was accorded the status of a distinct species. So attention turned to China and Japan, where it had been reported that trees were also afflicted, but survived the blight.

Frank Meyer had the wonderful job title of Agricultural Explorer of the Office of Foreign Seed and Plant Introduction of the United States Department of Agriculture.[28] He was traveling in northern China in 1913 when he received a letter from his boss, David Fairchild, in Washington, D.C., requesting that he search for symptoms of the disease among Chinese chestnuts. A few months later, Meyer's reply to Fairchild was published in *Science*, communicating an intimate picture of his commitment to his work that would be edited from scholarly journals today: "Here I am sitting in a Chinese inn in an old dilapidated town to the northeast of Peking . . . and have been busy for several days collecting specimens of this bad chestnut bark disease . . . It seems that this Chinese fungus is apparently the same as the one that kills off the chestnut trees in northeast America."[29] (I'll try this with my next journal submission: "Awaking from a sound sleep in my office this afternoon, I had a spectacular idea . . .") Meyer went on to note that although the fungus produced the same kinds of lesions that were seen in the United States and destroyed whole branches of the Chinese chestnut,

Castanea mollissima, the trees survived their illness. Meyer wasn't trained as a plant pathologist, but his observational skills were impeccable. He sent a sample of infected bark back to the USDA offices along with a box of nuts, suggesting that the Chinese trees might be immune to the blight. Back in Washington, plant pathologists were successful in growing the fungus from Meyer's specimens in the laboratory and confirmed that the same species of fungus was responsible for the infections in Asia and North America. They then deliberately infected American chestnut trees with the cultures to confirm their identification. Within a week, all of the inoculated trees developed symptoms of chestnut blight.

These crucial experiments followed the logic of the microbiologist Robert Koch, whose postulates, first detailed in the 1880s, remain central to the study of infectious disease. To determine the cause of a particular disease, the investigator begins by extricating the germ from infected tissues and growing it in isolation, or in "pure culture." Once the purported disease agent has multiplied, it is injected into a healthy host. If the disease is a human disease, a mouse or a monkey or some other unfortunate animal receives the injection. The investigator then sits and waits until the animal becomes sick and, if all goes well, the disease symptoms are similar to those in the original case. The monkey shivers in its cage and dies a horrible death, while the microbiologist considers his or her promotional raise over a nice cup of coffee. To complete Koch's recipe, the disease agent must be found in the lab animal. If all of these conditions are met, one can be confident that a causal connection has been discovered. In the study of HIV, the inability to complete Koch's postulates was a problem for researchers in the 1980s who were working to test the connection between the virus and damage to the human immune system. Lab animals were imperfect models for this inquiry because they didn't develop the symptoms of AIDS when they were injected with the virus. The case was clinched, however, by studying health care workers who developed AIDS after accidental exposure to the virus and patients who acquired the virus via blood transfusions.

On his way home from China in 1915, suffering from nervous exhaustion and depressed by the news of the continuing war in Europe, Meyer made a stopover in Japan. Pursuing a tip from a leading European plant pathologist named Johanna Westerdijk, who said she had seen the fungus in Japan, Fairchild had asked him to look for the blight in Nikko. Meyer discovered blighted trees in the hills around Nikko, and again, a couple of days later found the fungus on wild and cultivated trees in Yokohama.[30] Although the fungus was widespread, Meyer said that the disease was not very destructive. The trees displayed a level of resistance that had never been seen in the American chestnut.

Working a century after the appearance of the blight, Sandra Anagnostakis, at the Connecticut Agricultural Experiment Station in New Haven, has uncovered a compelling picture of the genesis of the disease.[31] Japanese chestnuts were first imported in 1876 by a nurseryman in the borough of Flushing, New York, just a few miles from the Bronx Zoo. Some of the original imports still grow in Connecticut. There are other records of sapling and seed imports in the nineteenth century, and hybrids between American and Japanese chestnuts were sold through mail order seed catalogs in the 1880s. By the turn of the century, the tree was growing throughout the East Coast. Chinese chestnuts didn't arrive in America until much later, around 1900, which argues against this species as the culprit for the earliest outbreaks.[32] Anecdotal reports suggest that trees were infected in Delaware, Virginia, and Pennsylvania a few years before the discovery of the disease in the Bronx Zoo, lending further credence to the idea that the disease was introduced on the Japanese chestnuts in the late nineteenth century.

Before I move on with this story, I feel compelled to mention the fate of Meyer, whom we left collecting in Yokohama. After a few months back in the States, Meyer made his will and returned to China for a fourth expedition. In 1918, the 42-year-old explorer disappeared from a riverboat on the Yangtze. His swollen body was found a few days later. An inquiry failed to establish how he slipped over the side rail, but a history of depression coupled with the bleakness of his later correspondence suggest that he jumped. During his time in Asia, Meyer had collected an astonishing 2,500 plant varieties that he thought had some potential as introductions to America. These included fruit trees and shade trees, drought-resistant elms (that were used to reduce soil erosion after the Dust Bowl), Chinese cabbage, alfalfa, bamboo, roses, disease-resistant spinach, the Meyer lemon (used to produce frozen lemon juice), and the soybean. Meyer collected 42 varieties of soybean and was enthusiastic about the Asian tofu industry decades before this slippery product was accepted in the American marketplace.

A century after the appearance of the blight in the Bronx, forests dominated by chestnuts have been replaced with forests filled with maples and oaks. Despite the deforestation to make way for new housing developments,

Walmarts, and Home Depots, the reversion of farmland to woodland has enlarged the green canopy in the eastern half of the country. A handful of apparently healthy chestnuts are dotted throughout the original range of the species. These plants are of great interest to researchers because there is a slim chance that they may have acquired some resistance to the fungus. Unfortunately, the survivors seem to owe their existence to quirks of timing and geography rather than any innate antifungal machinery. Planted after the blight epidemic, in areas where farmers cleansed the land of chestnut stumps, they grew up in blight-free isolation. Think about returning to your birthplace after an outbreak of Ebola: you have a good chance of surviving as long as all the corpses have been taken away. In the case of the rare chestnuts, however, the plague isn't likely to be very far away, and it will probably find them long before they become giants. This is because *Cryphonectria* kills every living cell in its victim's aerial tissues but stays clear of the root system. This mixed blessing allows the crippled plant to sprout new stems each spring that leaf-out and form loose baskets around the old stumps.[33] The regenerated stems soon succumb to the fungus. In this way, a once magnificent forest tree has been reduced to a shrub that is subjected to a continual cycle of damage and renewal. The chestnut and its blight are the botanical analogue of Prometheus and his liver-feasting eagle. If you subscribe to the glass-half-full philosophy of life, I suppose you could argue that after a century the chestnut and the fungus have established a balance that ensures that the genes of both organisms are here for the long haul. Since I belong to the "glass is empty, shattered, and I just stepped in the broken glass" school of thought, I think this viewpoint is no more sensible than arguing for the sanctity of a brain-dead patient on life support. At the Pennsylvania conference in 1912, Delaware's Agriculture Secretary, Wesley Webb said that the only way to eliminate the disease would be to destroy every tree. *Cryphonectria* was trying very hard to do this.

Planting chestnuts outside the tree's nineteenth century range proved the best way to avoid the fungus. Many of the biggest chestnuts now grow in a 60-acre stand of 2,500 trees in West Salem, Wisconsin. Established outside the natural range of the tree in the late 1800s, this population escaped the blight for a century. Since the first cases of disease were spotted in 1987, the grove has served as a living laboratory for researchers fighting the fungus. But despite heroic efforts to stall the blight, it seems likely that this Wisconsin stand is destined to become a chestnut morgue before long.

Healthy trees still grow in western states where the fungus was halted by the paucity of susceptible hosts west of the Mississippi and by the physical barriers offered by the Rockies and Sierras. Nobody knows how long these trees will escape the blight.

Optimists have always believed in a brighter future for the American chestnut. In the 1970s, chestnuts grown from seeds irradiated at the National Laboratories in Los Alamos, New Mexico, were planted across the country. The hope was that the chromosome damage caused by gamma rays would produce mutant trees that would resist the disease. Ohioan Bob Evans volunteered to plant some of them on his property. If you live within the original range of the chestnut, Bob Evans is a familiar name. It is written in big red letters on signs along highways, advertising 587 restaurants in 19 states. Since 1948, Bob Evans has been transforming pigs into breakfast meats and his bank account into a fortune. Unlike many others in the pig business, Bob is a dedicated environmentalist and has been honored on three occasions by the National Wildlife Federation. I visited him in the summer of 2004 to look at the 30-year-old chestnuts on his property in Rio Grande, Ohio. The trees were planted on a ridge above the idyllic farmland surrounding Bob's childhood home. Farm manager Ray McKinniss attempted to show me Chestnut Ridge from his SUV, but heavy rain had turned the track into soup. Thwarted by mother nature, Ray headed off for lunch, leaving me to find the trees on foot. Battling a powerful urge to escape the stabbing mouthparts of myriad invertebrates, I trudged miserably up the track, deciding that I would rename the ridge "West Nile Virus Park." Shadowed by healthy vegetation, a line of planted chestnuts materialized through my steamed glasses as splashes of large and unmistakable sawtooth leaves. The "trees" were terribly blighted: shrubs of mostly dead sticks. I took a few photographs, collected samples of infected bark, and ran back to civilization.

Bob was gracious enough to chauffeur me around the area in the afternoon. This part of the Ohio—Gallia County—has spectacular scenery; a gorgeous mix of well-groomed farmland and forest that merges with the Wayne State National Forest to the south. He showed me several locations where he had planted the saplings from Los Alamos. All the trees showed disease symptoms. The radiation hadn't created any mutants capable of rebuffing the fungus.

Radiation is one of several probably futile approaches to bringing back the big trees. Researchers at the State University of New York at Syracuse have been trying to combat the blight by creating a hybrid between chestnut trees and frogs. This isn't quite as ridiculous as it sounds. Rather than creating web-footed trees, the SUNY biologists hope to incorporate animal genes that confer resistance to fungal infections into the chestnuts.[34] Never mind that expressing frog genes in a tree in a way that would debilitate a fungus is more difficult than trying to split the atom with a hair dryer; never mind that chestnut blight doesn't affect frogs nor any other animal;[35] never mind a lot of other things that the researchers must have glossed-over with sufficient aplomb to obtain funding. (And never mind the fact that I spent a couple of years trying to get frogs to express genes from a fungus, and so I should be sitting very quietly in my glass house.)

A more sensible approach to controlling the disease developed from work in Europe. Within a few years of the blight's assault in the United States, the fungus had made its way across the Atlantic. The European chestnut is a different species, *Castanea sativa*, and though it is far smaller than the American version, it is highly valued as a nut producer.[36] First appearing in Genoa in 1938, chestnut blight spread to Spain, France, Switzerland, Greece, and Turkey. Plant pathologist Antonio Biraghi was among the first to study the disease in Italy, and echoed Hermann Merkel's struggles in the Bronx almost half a century earlier when he recommended cutting the diseased trees. The European trees survived infection for longer than their transatlantic cousins, but things looked very bleak. Then, in the 1950s, Italian trees that had been severely damaged by the fungus showed signs of recovery.[37] Cankers had healed and the fungus was confined to the outer bark, limiting damage to the plant's plumbing. In the 1960s, the French mycologist Jean Grente discovered that the blight fungus isolated from these trees was not as virulent as the American strains.[38] The identification of these hypovirulent strains was followed by a very significant finding. When active cankers were inoculated with this type of fungus, they began to heal. It seemed that colonies of the virulent fungal strain were weakened by contact with their hypovirulent relatives. Accumulating evidence showed that the hypovirulent strains carry a virus, called the hypovirus, with antifungal properties coded in two strands of ribonucleic acid. The virulent and hypovirulent strains were fusing with one another in cankers to create a hybrid fungus that was less aggressive. Inoculating

American chestnuts with the hypovirulent strains controlled the development of existing cankers, which was the first sign of promise for chestnuts since 1904.[39] But the investigators' excitement soon yielded to disappointment when fresh cankers developed on untreated tree limbs and eventually killed them. One of the problems is that the virus is transmitted by the asexual conidia. This means that an active colony that is growing without wreaking much damage nevertheless produces masses of wind-dispersed virulent spores that can attack and kill other trees.[40]

Hopes are now pinned on a more traditional program of breeding of proven value in the development of disease-resistant crop plants. Chinese chestnut trees rarely grow much larger than apple trees, and never grow into the kind of tall forest trees that once populated American forests. On the plus side, as discovered by Frank Meyer, Chinese chestnuts show a high degree of resistance to the blight fungus. When they are crossed with American chestnuts, the resulting hybrids inherit some of this resistance—enough, perhaps, to grow into large forest trees. The breeding program calls for repeated backcrossing of the offspring to American trees, to dilute-out the Chinese genes without losing the blight resistance. Scientists are now testing the resistance of trees that are more than 90 percent American. The effort is supported by The American Chestnut Foundation, based in Bennington, Vermont,[41] and involves scientists and volunteers across the country. My colleague, Carolyn Keiffer, at Miami University, has been instrumental in studying the resistance of hybrid trees. She and her students have planted thousands of saplings in Ohio and have deliberately infected many of these with cultures of the fungus to test their resilience. Although the tree that was devastated by the fungus may never return to American deciduous forests, perhaps one of the hybrids will repopulate the woods with a slightly Asian giant.

It is worth asking why anyone should bother about the American chestnut. After all, other trees are thriving in all of the locations where the chestnuts died. Here are a few answers. The chestnut might be valuable in wildlife restoration projects. Keiffer argues that disease-resistant hybrids will thrive on land damaged by strip mining, and there is plenty of it in the Appalachians. Chestnuts produce fruit within a few years, much faster than walnuts and other tree species. The nuts would offer a fantastic food source for animals, helping to revitalize these wounded lands. Rot-resistant timber would be a boon for the construction industry, supplying a new material for decking and tough roof shingles. Surprisingly, many foresters actually

Fig. 1.6   Loggers sitting in cut at the base of an enormous American chestnut tree. Photograph courtesy of The American Chestnut Foundation.

oppose the reintroduction of the tree, favoring the contemporary dominance of oaks in places like the Smokies over the fast-growing chestnut. I don't think they have much to worry about.

The devastating effect of *Cryphonectria parasitica* upon our forests is usually ascribed to the fact of its introduction from Asia. Just as European explorers decimated native societies by introducing venereal diseases, smallpox, influenza, and other nasties to which the victims had little resistance, disease-causing fungi can thrive on the powerless prey they encounter when they escape from their homelands. It is very unlikely that chestnut blight would have taken hold in America without the aid of human importation. But there is a bit more to this story. Look at the photograph of the woodsmen dwarfed by the redwood of the east (fig. 1.6). This is an execution-day mugshot of a giant that the lumberjacks are about to topple. The photograph was taken in the nineteenth century, years before the discovery of the blight. Virgin chestnut forests had disappeared long before the fungus showed up. The vast majority of the trees that the fungus killed were relatively young, products of the secondary-growth forest that greened the eastern states after the biological holocaust wrought by the westward migration of Europeans.[42] As I explained earlier, chestnuts were actually more plentiful after the orgy of logging and burning because they outgrew other hardwoods. In some areas, human disturbance had created a monoculture of chestnuts that welcomed the blight fungus. It's also possible that the young trees weren't as hardy as the inhabitants of the ancient forest, and they were probably less diverse genetically. The age structure of the forests had also been upset by forcing complex communities of organisms to rebuild themselves from scratch.

In a continent so disturbed by humanity, the virtual extinction of the American chestnut was probably inevitable. It ranks as a one of the worst biological catastrophes produced by a collaboration between humans and fungi. But there are plenty of others I'd like to tell you about.

CHAPTER 2

A Farewell to Elms

Without hitches at airports, I can get from my home in Ohio to the tiny English village where my parents retired in less than 14 hours. Drayton St Leonard in Oxfordshire is a lovely place. Despite soaring house prices and changing demographics (more retirees and commuters today than farmers), the population has been stable at around 300 since 1800.[1] A few homes have been built in the center of the village, and barns have been converted, but the village remains within its medieval boundary, surrounded by farm fields, the black tower of its church visible for miles across the flat valley of the River Thames. Sipping tea in a blissful garden, the music of insects and the church bell chiming the hour, one is liable to forget about the inexorable fungal assault upon the landscape. Fortunately, for the purpose of this book, the history of the village tells that this place, too, is rotten: All of Drayton's magnificent elms are gone, strangled by Dutch elm disease.

The church and schoolhouse were once dwarfed by enormous trees, and long-term residents recall the avenues of elms along the web of roads around the village. The last of the elms died in the 1970s, victims of the Dutch elm fungus, *Ophiostoma ulmi*. The loss of the elms in an English village is a trivial story. I doubt that anyone in Drayton St Leonard lost much sleep over the removal of the dying trees (before they could fall on the school). The villagers sighed, shrugged their shoulders, and got on with their lives. But multiply the effect of the fungus in Drayton across all of Europe and North America, and chestnut blight meets a serious rival for the award of Nastiest Fungal Disease of All Time.

In terms of raw numbers, Dutch elm disease has not killed as many trees as chestnut blight, simply because there were never as many elms as chestnuts. But the fungus continues to infect trees in the Northern Hemisphere

25

in the twenty-first century, and its victims already number more than 100 million. Whereas *Cryphonectria* limits itself to American chestnut and a couple of close relatives, *Ophiostoma* can kill almost any species of elm, or *Ulmus*. There are a lot of species of elm. In Western Europe, the English elm, *Ulmus procera*, and wych elm, *Ulmus glabra*, were the most important casualties (fig. 2.1). The English elm was a massive tree, with record specimens measuring 46 meters in height with a 2-meter diameter shadowing even the American chestnut. This "majestic plant," as the essayist John Evelyn called it, was a dominant feature of the British landscape and was ubiquitous in hedgerows. The wych elm was somewhat shorter than the English elm and spread its branches into a dome-shaped crown. Besides these species, the pathogen killed regional varieties of the smooth-leaved elm, *Ulmus minor*, myriad hybrids between naturally occurring elm species, and horticultural varieties. Across the Atlantic, the American or white elm, *Ulmus americana*, was the common species and offered minimal resistance to the disease. The American elm overlapped the natural range of the chestnut, but also grew farther west, spreading throughout the Dakotas, Nebraska, Kansas, Oklahoma, and the northeastern third of Texas. The diversity of host trees, coupled with their fantastically wide distribution, resulted in a disease history that is far more convoluted than the single, exuberant southwesterly drive of the chestnut blight fungus.

It happened like this. In the nineteenth century, elms in city parks in Britain and mainland Europe succumbed to a disease that some contemporary experts believe was Dutch elm.[2] If these deaths were an early manifestation of the disease, then the pathogen might have been a long-term resident of Europe rather than an Asian immigrant like the chestnut blight. The first well-documented epidemic of Dutch elm disease developed in northwestern Europe at the end of the First World War.[3] The name of the disease refers to its discovery in the Netherlands.

A remarkable feature of the early research on Dutch elm disease is that the key investigators were women.[4] Dutch elm disease was recognized as an "unknown disease among the elms" by Dina Spierenburg, a staff member at the Institute for Phytopathology in Wageningen. Yellowing leaves and drooping shoots were the first symptoms of infection, followed by widespread staining and occlusion of vessels that first debilitated, and then killed, the tree. The fungus that caused the disease was identified by a graduate student, Marie Schwarz, under the direction of Johanna Westerdijk (fig. 2.2)

Fig. 2.1 Leaves, flowers, and fruit of English elm, *Ulmus procera*. From F. A. Michaux, *The North American Sylva; or, A Description of the Forest Trees of the United States, Canada, and Nova Scotia. Considered Particularly With Respect to their Use in the Arts and their Introduction into Commerce. To Which is Added A Description of the European Forest Trees*. English translation, vol. 3 (Philadelphia: D. Rice and A. N. Hart, 1857).

at the Willie Commelin Scholten Phytopathological Laboratory in Baarn. Another of Westerdijk's students, Christine Buisman, resolved the sexual part of the organism's life cycle, and other young Dutch women are credited with establishing the facts about the development of the disease and were the first to examine treatment methods. In contrast, all of the major players in the story of chestnut blight were men: Hermann Merkel, the forester who discovered the disease; William Murrill, who identified the fungus; the majority of delegates at the 1912 blight conference in Pennsylvania, and so on. As I'll explain later, the sex of the Dutch elm investigators played an important role in the study of the epidemic.

A number of explanations for the disease (some verging on delusional) were put forward after Spierenburg alerted the scientific community to the problem. In France, drought was seen as a possible cause, and others tried to implicate the poison gas deployed on the European battlefields. Various fungi were blamed, and a German scientist isolated a bacterium from infected elm tissue (that was never found again), before the true culprit was

Fig. 2.2 Johanna Westerdijk, founder of a school of research on Dutch elm disease in the Netherlands. From G. C. Ainsworth, *An Introduction to the History of Mycology* (Cambridge, UK: Cambridge University Press, 1976), with permission.

discovered. To isolate the fungus responsible for the disease, Marie Schwarz disinfected the surface of small pieces of diseased wood, washed them with water, and placed them onto an agar medium that contained an extract from cherries. Cherry extract is acidic and inhibits the growth of bacteria and yeasts, allowing any slower growing fungi to emerge from the wood. Within a few days, a white colony spread in a ring around the wood samples and produced stalks that bore groups of tiny spores held together in mucilage. Schwartz wrote that, "These small spore-heads appear, even under a hand lens, as clear, glistening droplets on the snow-white mycelium."[5] This was the asexual stage of the Dutch elm pathogen that she christened *Graphium ulmi*. Following the same logic applied by the chestnut blight investigators, she then infected healthy elms with her cultures to test whether she had found the causal agent. The wood of experimental trees inoculated with the *Graphium* turned dark brown, just as it did in cases of natural infection. Schwarz had clinched the deal.

Now we come to the significance of gender. In hindsight it is clear that the 24-year-old student had identified the cause of Dutch elm disease, but very few people believed her in the 1920s. How could a student, especially a young woman (poor little clog-dancing blonde girl in pigtails, etc.), have made such an important breakthrough? The weakness of the alternative explanations for the disease didn't seem to hold any sway among Schwarz's critics. After graduation, Schwarz returned to Indonesia (her birthplace) and continued her research on plant diseases. Meanwhile, the disease spread throughout Europe, qualifying Dutch elm as a pandemic. (The term pandemic is used to describe a disease that afflicts a larger geographical area than an epidemic. There isn't a sharp distinction between an epidemic and pandemic, but I'll try to clarify things with an example. Many years ago, teenagers in the United States were enthralled by a dopey adolescent named Britney Spears. An *epidemic* of interest developed. Her fan-base spread to Europe, Japan, Afghanistan, and so on, creating pandemonium and a Britney *pandemic*. Mercifully, however, interest in Britney subsided, and no more than a few sad-faced citizens can recall her vacant expression today.)

Although Schwarz's discovery didn't suggest an effective treatment for Dutch elm, the fact that researchers ignored the young woman's work certainly frustrated progress in understanding the nature of the problem. Christine Buisman had already completed her doctoral research on root rot of lilies when Westerdijk assigned her to repeat Schwarz's experiments. Westerdijk saw no other way to settle the deadlock between supporters of the *Graphium* indictment and those who continued to dismiss Schwarz. Buisman swiftly verified that the fungus caused the disease and further demonstrated that the experimental infection of elms only reproduced the symptoms of wilting if the trees were inoculated in early summer.[6] This was an important finding because Schwarz hadn't been able to reproduce the whole suite of Dutch elm symptoms in her experiments. Buisman also discovered the sexual stage of the fungus, which resembled the perithecia of the chestnut blight fungus. The naysayers were silenced: not poison gas, not an elusive bacterium, but a fungus that was renamed *Ophiostoma ulmi*.

Marie Schwarz married in 1926, retired from science, and had two sons. Her move to Indonesia had isolated her from the controversy about the cause of Dutch elm disease. But during a visit to the Netherlands in 1931 her colleagues invited her to a conference in Belgium where Christine Buisman presented the supporting evidence that ended the dispute. The subsequent lives of both women followed interesting and tragic paths. After suffering internment by the Japanese in the second war, during which her husband died, Schwarz returned to Baarn with the children. She spent the rest of her life studying black-pigmented molds. Buisman was the first to identify Dutch elm disease in the United States in 1930. A Cleveland arborist had found an elm with wilting foliage and, after felling the tree, noted dark streaks in the infected wood.[7] Hearing that Buisman was visiting Harvard's Arnold Arboretum, he had sent her an infected twig to diagnose the problem.[8] Back in Europe, Buisman began work on the selection of disease-resistant elms, consolidating her international reputation as an expert on tree diseases, but she died in an Amsterdam hospital in 1935 during cancer surgery. The proverb stating that "Every elm has its man" is utterly opaque, but is illuminated by revision: "Every elm had its woman."

ða

The utter flatness of the Netherlands landscape was exposed in the wake of the disease. More than one million Dutch elms, a hybrid tree designated *Ulmus × hollandica*, had lined the roads and dykes; the fungus killed 700,000 of them. But the reason that the Netherlands became the center for research on the pandemic was the leadership of Johanna Westerdijk. After all, by the end of the First World War, elms were also dying throughout Belgium, France, and Germany, where there were plenty of homegrown scientists. The first case in Britain was identified on a golf course in the village of Totteridge in Hertfordshire in 1927.[9] By 1930, the disease had spread throughout southern England. Estimates suggest that between 10 and 20 percent of the elms were killed in this first outbreak.

The speed of the elm pathogen's spread across Europe has often been cited as an alarming feature of the disease, but this speaks more to confusion about the behavior of microorganisms than anything spectacular about this particular fungus. The incomprehensibly vast numbers of microscopic spores, coupled with their insistent mobility, ensures that many nasty diseases can, and often do, spread over entire continents in a few years. In the previous chapter I discussed the rapidity of North American colonization by the chestnut pathogen. In that case, the spores were spread both by wind and by passive attachment to any bird or insect that happened to land on an infected tree. Dutch elm fungus is more finicky. Like the chestnut fungus, it produces two spore types: an asexual conidium and a sexual ascospore. *Ophiostoma* uses both spore types to spread from elm to

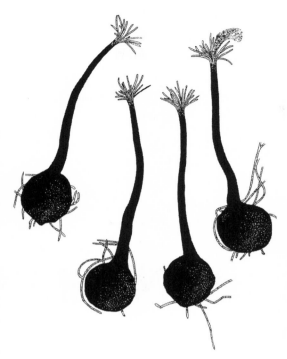

Fig.   2.3   Spore-extruding fruiting bodies (perithecia) of *Ophiostoma ulmi*, the Dutch elm fungus, formed within beetle galleries. From C. J. Buisman, *Tijdschrift over Plant-enziekten* 38 (1932).

elm, but unlike the spores of the chestnut pathogen, they aren't carried by wind. As Schwarz described them, the asexual conidia are produced in sticky blobs at the tips of millimeter-high stalks. The sexual ascospores are extruded through the long necks of the fruiting bodies, or perithecia (fig. 2.3), of the fungus, but also remain clumped together. Sticky spores are not adapted for dispersal in wind, and an even greater impediment to being blown away is the fact that they are formed inside infected elm trees. The way the spores moved from tree to tree was discovered in 1934, when another Dutch student who worked with Christine Buisman, this time a young man named J. J. Fransen, demonstrated that the fungus was delivered by bark beetles that burrow in elms. The fungus produces its sticky spores on the walls of the beetle galleries, so that beetles emerging from diseased trees are coated with infectious cells.

The adult insects are attractive critters, though their beauty isn't apparent until they are magnified (fig. 2.4). With a length of 5 millimeters or so, the large elm bark beetle, *Scolytus scolytus*, is twice the size of the small elm bark beetle, *Scolytus multistriatus*. Both carry the spores on their bristly,

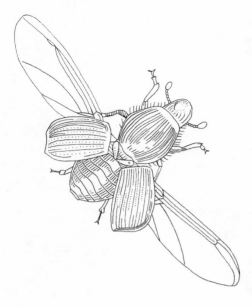

Fig. 2.4   The elm bark beetle *Scolytus scolytus*. Drawing by Thomas J. Cobbe, courtesy of the Willard Sherman Turrell Herbarium, Miami University, Oxford, OH.

black bodies. This is unfortunate for elms because the highly infectious beetles fly around sniffing the air for the odor of an elmcrotch. Elm crotches, since you ask, are the angles where twigs meet larger stems. Their odor results from a mixture of chemicals including vanillin, that has the smell of "sweet dry vanilla chocolate," and syringaldehyde, which smells like "fresh grass."[10] The insect response to these temptations is comparable to my reaction to the aroma of roasting coffee beans: though flightless, I would, if necessary, chew through bark to get my morning buzz.

Crotches and small stems are the sites where beetles invade healthy trees. The female beetles munch their way through the bark and into the sapwood beneath, leaving an engraving of their path through the tissues of the elm. Male beetles are attracted by pheromones released by the females. After mating, the impregnated female bores an upright shaft and lays about 70 white eggs in an alternating pattern on either side. When the eggs hatch, the larvae chew their own tunnels at right angles to the mother's burrow. The result is an elaborate gallery of tunnels that form the distinctive signature of elm bark beetle infestation, and, invariably, Dutch elm disease. Imported by the beetles, the fungus burgeons in the humid galleries and extends its hyphae into the surrounding wood. After the beetle larvae pupate and metamorphose into adults, the insects escape through the bark and carry a coating of spores. This is how Dutch elm disease spread in Britain. But how did it arrive in Ohio where Christine Buisman identified it in 1930?

The obvious threat to American trees posed by the European epidemic was recognized as soon as the scope of Dutch elm disease in the Netherlands became apparent, and imported nursery stock (ornamental elm saplings) had been screened on arrival in American ports. Unfortunately, the inspectors "missed the wood for the trees." (Apologies for the cliche, but it works so admirably in this context, as you'll see.) The saplings were turned away, but bark beetles carrying the fungus arrived on elm burl logs, also known as Carpathian elms, imported for furniture manufacture. Burls are massive tumorous lumps that ornament older specimens of many tree species; sliced into veneers, their bird's-eye grain makes gorgeous swirling patterns that are prized by cabinet makers. Shipments arrived at the ports of Hoboken, Baltimore, and Norfolk, and were transported by railroad and on river barges to Cincinnati, Louisville, and other Midwestern centers of furniture manufacture.[11] The diseased trees discovered in Ohio were probably infected by beetles that arrived on one of the 100 shipments documented between 1926 and 1930. The connection between the burl logs and Dutch elm disease was proven in 1933 when *Scolytus* beetles carrying *Ophiostoma* spores were identified in a shipment of logs from France that arrived in New York. The connection between burl logs and the furniture industry and Dutch elm disease was strengthened by the discovery of infected trees next to piers in Norfolk, Virginia, where logs had been unloaded in 1934, and in Indianapolis next to a veneer factory. A quarantine directive regulating the importation of elm logs took effect in 1933, but by then the disease had spread over an astonishing range. Investigators found the beetles and the fungus along thousands of miles of railroad lines that had carried the logs.

One of the European carriers of the fungus, *Scolytus multistriatus*, had arrived in North America well ahead of the disease. The beetle was discovered in Massachusetts in 1904 (the same year that Hermann Merckel encountered chestnut blight in the Bronx), where it munched harmlessly on tree bark. Trees in the scenic area along the Ohio River between East Liverpool, Ohio, to Evansville, Indiana, had been infested with European bark beetles for years before the appearance of Dutch elm disease. Presumably, burl logs replete with insect-infested bark had been showering the countryside with beetles for years. In addition to the established populations of imported carriers, the fungus would meet another confederate in the form of a native bark beetle called *Hylurgopinus rufipes*. The stage was set for a botanical massacre.

After the first cases in Cleveland and Cincinnati in 1930, four more infected elms were found in Cleveland in 1931. Faced with the recent history of Dutch elm disease in Europe, and experiencing the third decade of chestnut blight, American foresters realized that efforts to save the elms would be futile. In 1932, no diseased trees were found in Ohio, but the respite was brief.[12] After the first outbreaks in Ohio, cases of Dutch elm appeared in New York City and the surrounding area, with hundreds of infections in 1933. It was, indeed, "all over but the shouting." In 1942, 65,000 diseased elms were removed from an area of 31,000 square kilometers, equivalent to two diseased trees in every square kilometer.[13] Compared with the figures for chestnut blight this doesn't sound particularly alarming, but each of the victims of the new disease was a valuable shade tree growing in a city. So although chestnut blight was a more ferocious problem (which is a little like trying to rank Hitler versus Stalin), Dutch elm disease injured the spirits of far more citizens than chestnut blight. Elms were also killed in forests, of course, millions of them, but the infamy of Dutch elm draws from its recognition as the worst tree disease to have affected the urban environment.

I walked along Elm Street in Cincinnati this morning and tried to picture the stately trees that died 60 years ago. Today, the only elms on Elm Street are a line of scrappy Chinese elms, *Ulmus parvifolia*, outside the Convention Center. There are countless examples of sad stories about elms in the United States, just like the strangled trees in Drayton St Leonard. Here's one of them. In 1816, delegates met in Corydon, Indiana, to draft the state constitution. They worked on this sacred document under a huge elm that became known as the Constitution Elm. A century later the tree was killed by the disease and its crown was cut and replaced with an octagonal canopy. The memorial looked like an oversized patio umbrella.

With the aim of creating a disease-free barrier that would halt the spread of the fungus-carrying beetles, the federal government instituted an eradication program in the 1930s that called for the destruction of diseased trees. (This should sound familiar. Remember the futility of the Chestnut Tree Blight Commission?) Some of the plant pathologists who had played starring roles in the fight against chestnut blight joined the efforts to stem this second epidemic. Haven Metcalf, chief of the USDA's Laboratory of Forest Pathology, who had investigated the Asian origins of the chestnut pathogen, visited Europe to examine the scale of Dutch elm disease. He was one of several pathologists who predicted the arrival of the disease in North America. Another USDA scientist, Curtis May, set up a laboratory in a former speakeasy in Morristown, New Jersey, to study Dutch elm. May and colleagues isolated the fungus at the bar, with, I presume, the ghosts of inebriated flappers disturbing their work every now and then.

Federal funding for tree eradication ceased during the Second World War, and by then the disease had spread as far north as Quebec. At around this time, the westward advance of the epidemic caught up with a second lethal disease of elms called elm yellows, or elm phloem necrosis. This is caused by an unusual microorganism called a mollicute which is related to bacteria. Mollicutes get inside the bark of elms with the help of an insect vector such as the white-banded elm hopper, destroy the phloem, and kill the tree with a list of symptoms that is similar to those of Dutch elm disease. There are a couple of distinguishing features of elm yellows. If you cut and peel the bark with a knife, you'll see that the diseased tissue has a vivid lemon yellow color. In addition, sniff the cut bark and you may detect the odor of wintergreen breath freshener due to the emission of methyl salicylate from the dying tissue. Cases of elm yellows have a spotty distribution, and the mollicute tends to flare-up in one area without spreading very far. But the availability of trees killed by mollicutes west of Pennsylvania offered a perfect breeding ground for the bark beetles arriving in the 1940s. Things were looking increasingly bleak for *Ulmus*.

The campus of Miami University in Oxford, Ohio, was clobbered by Dutch elm disease. The city was carved from primeval forest after the Miami Indians had been forced from their hunting grounds and sent off to Cincinnati and river passage to permanent exile by the white settlers. After the Indians' departure, there was a great deal of deforestation; in fact, almost nothing of the original forest was left standing by the end of the nineteenth century. The settlers reversed the proportion of woodland to open space, taking an Eden and creating its photographic negative. But after the university was established in 1809, some vegetable matter seemed like a good idea for the campus. Yale, already in its second century, was in New Haven, "the City of Elms," and Harvard and Princeton had a few of their own trees. Clearly, red-brick buildings surrounded by hundreds of square miles of burning stumps and eroded soil would have attracted few but the worst scholars. To this end, a local dentist named George W. Keely collected seedlings of oaks, elms, and maples from the remaining woodland and planted them all over Oxford.[14] The outcome was impressive: for the first half of the twentieth century, the campus was shaded by 1,800 elms and the city streets were well-elmed, too. Then entered the bark beetles, and, shortly thereafter, the work crews to dispose of 1,800 dead trees.

Elms had been viewed as ideal urban trees by city planners. When the University of Illinois was established in Urbana after the Civil War, the board of trustees transformed the treeless prairie by planting elms—one of them describing the American elm as "the tree that like a fountain rises." Dutch elm disease arrived in Urbana in 1951 and killed more than 2,000 trees in a decade. Fortunately, other tree species had been planted. The planners of Moline, Illinois, 180 miles to the northwest, lacked the same aesthetic: elms were the only trees in the city. As Richard Wolkomir wrote, "Towns became naked."[15]

Among all of the elm species and hybrids, arborists regard the American elm as exceptionally vulnerable to attack by *Ophiostoma*. Beyond a few clues, the reason for differences in susceptibility among elms isn't clear, and similar haziness attends scientists' understanding of much of the specificity that other tree pathogens show for their hosts. Why, for example, doesn't the Dutch elm fungus infect maples? After all, there are lots of maples in North America and in Europe (though they are called sycamores in Europe),[16] and there are plenty of calories locked away in their bark and wood. The answer lies in multiple features of the biology of pathogenic fungi and trees. The arrangement of cells in the wood of elms and maples is different, and the chemical structure of these tissues is also distinctive. *Ophiostoma* has evolved the enzymes needed to decompose elm wood but, perhaps, these are ineffective against the corresponding chemical mixtures in maple trees. Like all plants, elms are engaged in a chemical arms race with the fungi and produce an array of antifungal compounds to impede colonization. At this point in the evolutionary history of elm trees, the Dutch fungus is winning. But the elm's defenses presumably work against other microorganisms: chestnut blight doesn't plague elms. Finally, by acting as couriers for their spores, woodland insects play important roles in determining which fungus afflicts which tree.

Like chestnut blight, Dutch elm disease kills trees by attacking their vascular supply. The fungus *Ophiostoma* penetrates the water-conducting xylem tissue more deeply than *Cryphonectria*, wounding the wood and leaving dark bands of blocked vessels.[17] The effect has been likened to arteriosclerosis. This is an instructive metaphor when one considers the lethal effect of scraps of plaque when they are shed from hardened blood vessels and circulate to the brain. In the Dutch elm version of embolism, the fungus produces budding yeast cells in the interior of large vessels that are transported through the fluid. These "bud cells" can establish new infection sites as they flow around the tree, and by sticking to the vessel walls they also interfere with water transport.

Like any pipe, wider vessels are more effective than narrower ones at moving lots of water. But this conductive efficiency comes at a price. The silvery columns of water flowing through larger vessels are prone to snapping. It is, perhaps, counterintuitive to think about water acting like snappable string, but water columns do act in this fashion because the constituent water molecules are attracted to one another and resist separation. This is one of the factors that allows trees to pull continuous columns of water all the way from their roots to their leaves. The tension that develops in the slender water columns as they are pulled upward inside vessels is phenomenal, and if this tension exceeds the cohesion between water molecules, the columns will snap apart. This is called cavitation, and it severs the water supply to the leaves. Cavitation, particularly in large xylem vessels, is responsible for the wilting symptoms of Dutch elm disease.

A few years ago, plant pathologists became very interested in a compound called cerato-ulmin secreted by the Dutch elm fungus. It was thought that cerato-ulmin acted as a toxin and was directly responsible for disrupting water transport. Experiments showed that cerato-ulmin was a type of hydrophobin, or water-repellant protein, that seemed ideally suited for blocking vessels. But a satisfying explanation isn't necessarily correct, and genetic experiments disproved this one. Canadian researchers created mutant strains of the fungus that produced none of the putative toxin but retained the unfortunate knack of murdering elms.[18] Vessel blockage in Dutch elm disease seems to be caused by the production of gummy material called tyloses by the infected tree, which is the same mechanism that leads to wilting in blighted chestnuts. Xylem blockage is useful if it enables the infected tree to seal-off vessels that contain fungal cells, but as the fungus spreads, this defense mechanism becomes increasingly dangerous as the tree is forced to shut down more and more of its water supply.[19]

Between 1930 and 1960, when Dutch elm disease was in full swing in the eastern United States, populations of elms across Europe seemed to be rebounding. Millions of the infected trees had died, but many others produced new growth and recovered. In Britain, Dutch elm disease "came to be regarded as an endemic problem of no great consequence."[20] Then, in the late 1960s, fresh outbreaks were reported in southern England. The disease was far worse this time. While the first epidemic killed between 10 and 20 percent of the trees, the second outbreak spared no elm. The Drayton St Leonard giants, and 11 million relatives—around half of the total elm population in the southern counties—were gone in a decade.[21] Pollution was advanced as a possible explanation for the increased susceptibility of the trees, but then it was recognized that the elms were infected by a new strain of *Ophiostoma*. The fungus had become more aggressive. The reappearance of the disease in Britain presented a tremendous challenge for a new generation of plant pathologists. Detective work by John Gibbs and Clive Brasier at a Forestry Commission research station (the Alice Holt Lodge in Surrey) revealed that the fungus had arrived on logs of rock elm imported from Canada.[22] Rock elm, *Ulmus thomasii*, produces a straight-grained and very hard timber that is used for boat building. One shipment from Toronto that was examined in Southampton was riddled with the fungus and its American beetle vector, *Hylurgopinus*.

The fury of the second outbreak of Dutch elm disease in Europe was due to the appearance of more aggressive strains, or "races," of the fungus.[23] The fungi that caused the European epidemic during the First World War are now referred to as the "nonaggressive" subgroup, which is a relative term because the fungus was sufficiently assertive to have changed the landscape in the Netherlands. Two races with enhanced virulence developed later in the twentieth century. The Eurasian version developed first and may have made its way to North America in the 1940s. Clive Brasier, the leading authority on the disease for the last 30 years, believes that the Eurasian race mutated on the highly susceptible American elms, becoming even more vicious.[24] Whether or not this is true, the North American aggressive race of *Ophiostoma* was introduced to Europe in the 1960s and found British trees easy prey. The detailed history of the pathogen is unresolved, but it is clear that bark beetles and human commerce collaborated to swirl an ever-changing pathogen around the Northern Hemisphere.

The transatlantic and transcontinental migrations of the disease represent one of the most interesting features of the disease, but where had *Ophiostoma* come from originally? Just as chestnut blight had been tracked to Asia, there was a lot of interest in the possibility of a Chinese origin for

Dutch elm disease. The Himalayas were considered a likely birthplace for the fungus, but recent research hasn't supported this hunch. An origin elsewhere in Asia remains a strong possibility, but some versions of the story of transcontinental transport are ludicrous. Let your imagination go wild for a moment. Make up a silly story about the introduction of Dutch elm disease to Europe in the twentieth century and you may have difficulty doing better than Edwin Butler's 1934 "wicker basket theory."[25] How about thousands of Chinese laborers, recruited to work behind the battlefields during the First World War, who marched through the Netherlands disseminating the disease from beetle-infested baskets? I quote from an article by Horsfall and Cowling published in 1978: "They moved their meagre belongings in wooden wicker baskets made from the tough fibrous wood of the Chinese elm. Some of the pieces carried bark and the vector beetles. Presumably the fungus escaped into the low countries."[26]

Truth is, of course, often stranger than fiction, but the wicker basket theory is stretching the bounds of sane speculation. John Gibbs dismissed the idea in a scholarly review of the evidence published in 1980, saying with great wit that "the wicker basket theory does not hold water!"[27] The British indeed transported 100,000 Chinese men to Europe between 1917 and 1918 to work behind the Allied lines (rather than digging trenches as some have said). But regardless of the biological diversity clinging to their belongings, it is important to remember that this mass human migration coincided with the appearance of the disease in the Netherlands. The idea would be more sensible if the laborers had arrived before the war and before the epidemic. I'm all in favor of blaming the Chinese, but in this instance they seem faultless.

One school of thought argues that *Ophiostoma* has been killing European elms for thousands of years and that it is an endemic pathogen that flares up from time to time. Studies of pollen records in Irish peat bogs show a sharp decrease in elms around 3100 B.C.[28] In earlier sediments, elm accounts for one in five pollen grains from trees; a few centimeters higher, and elm pollen crashes to 1 percent of the mixture. The same disappearance of elm is seen in pollen records elsewhere in northwestern Europe dated to this period, called the Atlantic to Pagan Transition. This pattern might have been due to catastrophic disease, although deforestation for agriculture would obviously have had the same effect. The later pollen record from Ireland supports the disease idea, because elm reappears for a while after 3100 B.C., then vanishes in 2400 B.C. when we know that panoptic deforestation of Ireland by

Neolithic farmers transformed the country into a treeless, boggy island. Because there is no direct evidence that elms were attacked by *Ophiostoma* 5,000 years ago, however, the value of the pollen record in interpreting the early history of Dutch elm disease is limited.

Evidence of a nineteenth-century outbreak is stronger. Under the pen-name Dendrophilus ("tree lover"), the *Philosophical Magazine and Journal* published a whimsical account of damaged elms in St. James's Park in London in 1823.[29] Bark was stripped from the "Lungs of London" with such violence that human vandalism was suspected, and "persons were employed to sit up during whole nights watching for the enemy." Recounting the story to his Victorian readers, Dendrophilus reported that minute bark beetles, rather than the suspected soldier's bayonets, were responsible for the damage. Some authors see this, and other stories of elm damage in Europe from the eighteenth and nineteenth centuries, and paintings of dead elms from the Italian Renaissance, as evidence of early outbreaks of Dutch elm disease.[30] But others have concluded that drought and beetles would have done the trick without the fungus.

Great efforts have been made to restore elms in American cities. As shade trees, American elms are (or were) hard to beat. A medicine man from South Dakota proposed curing diseased elms by injecting an ancient medicine of unspecified composition. The fee for this service in the city of St. Paul, Minnesota, was quoted at $1 million, but in the absence of a portfolio of successful cases, the city took a raincheck. Plant pathologist David French has documented a plethora of similarly far-fetched solutions.[31] A florist and allergist in Florida developed a decongestant liquid that would clear the blockages in the diseased xylem vessels. A dentist thought that mineral supplements would make trees healthier and disease resistant, and, in a similar vein, some other wizard proposed that elms needed an invigorating dose of ground carp and seaweed. But even sillier cures were proposed, including the broadcast of music and high frequency sounds, and, finally, the construction of pyramids in tree groves. Have you visited Giza? Bet you didn't see any diseased elms. Like most fungi, *Ophiostoma* is terrified of pyramids.

Moving far up the scale of human intelligence, a series of sensible but ineffective approaches to the Dutch elm problem have been proposed. Biological control involving a bacterium that inhibits the growth of many fungi is a reasonable idea, but it doesn't work. Using sexual lures (the insect pheromones I mentioned earlier) to capture bark beetles also sounds good, but this doesn't

CHAPTER 3

The Decaffeinator

I am not certain when my passing affection for coffee became a dependency, but I am sure of my place in the pantheon of caffeine addicts. Some evenings, I fix an espresso simply to enable my ascent up the stairs to bed. Without this final fix of the day, I will sleep, fully clothed, wherever I slump; after the gulp of bitter balm, teeth will be cleaned, clothes will be removed, but I'll still be asleep before I can count back from 10. This dependency on the coffee plant, *Coffea arabica*, is shared with the rust fungus *Hemileia vastatrix*, though I need the beans, and it eats the leaves (fig. 3.1). We are united in consumption and in competition.

The restorative powers of coffee are said to have been discovered by an Abyssinian goatherd named Kaldi, who noticed the invigorating effect of the red fruit on his cloven-hoofed charges. ("What a cartload of bollocks," I hear you cry, and I agree that this fable has no more basis in reality than a flying carpet.) It is presumed, with greater confidence, that early human use of the beans involved a travelers' snack of bean-encrusted animal fat, because this is still consumed in parts of Africa.[1] Today's chocolate-coated beans can be considered (though not by sensible persons) as a reminder of this refreshing Atkins-approved fat-ball. Concurring with the Kaldi story, wild arabica coffee is native to Ethiopia, and the southern region of Sidamo remains the source of some of the loveliest brews, including the wonderfully aromatic beans from Yirgacheffe. A few hundred kilometers to the north, in the Rift Valley, Lucy and other important hominid fossils have been unearthed, and this region is considered the likeliest birthplace of our species. I mention this, because it is interesting to muse, as other authors have, on the possibility that Lucy and her kin enjoyed coffee snacks millions of years before

Fig. 3.1 Leaves and berries of arabica coffee, *Coffea arabica*. From M. Buc'Hoz, *Dissertations sur L'Utilité, et les Bons et Mauvais Effets du Tabac, du Café, du Cacao, et du Thé, Ornées de Quatre Planches en Taille-Douche,* 2nd edition (Paris: L'Auteur, 1788).

Kaldi's goats. Humans and coffee have certainly enjoyed a mutually supportive relationship for thousands of years.[2]

Following our shared exodus from Africa, *Coffea* and humans have migrated together, bean in hand, around the planet's midriff. Today, 25 million coffee-farming families in 60 countries produce between 6 and 7 megatonnes of beans a year.[3] For comparison, wheat production is edging close to 600 megatonnes, which isn't surprising given that wheat is a staple food, and that nobody need drink coffee. But coffee is second only to oil among the most valuable lawfully traded commodities, and it accounts for $70 billion in consumer spending. From its Ethiopian homeland, coffee was traded across the Red Sea, to Yemen, and was embraced across the Muslim world. Europe's introduction to coffee came in the form of Turkish exports originating from the Yemeni port of Mocha in the sixteenth century. The Dutch were the first Europeans to grow coffee, beginning in Ceylon (Sri Lanka) in 1658, and later throughout the East Indies. Now, I need to keep you in Ceylon for a while, but I'll return to the continuing spread of coffee in a few pages.

Control of Ceylon was transferred to the British in 1796, and the country became a crown colony in 1802. Dutch resistance to the takeover was limited. They had been preoccupied with homeland issues since their invasion by revolutionary France in 1795, and the British had been unofficially colonizing the island for years. According to many British sources, the "old colonels" (my favorite blanket term for the colonials) behaved very decently in Ceylon, abolishing slavery, establishing an effective judicial system, and extending the civilizing influences of history's greatest nation. Ceylon's forests were also civilized and encouraged to produce cinnamon, pepper, sugarcane, cotton, and lots and lots of coffee. The native Kandyan Sinhalese of the hill country remember things differently. Their version of history includes the loss of tribal lands, a broken convention, the destruction of villages, and slaughter of anyone offering the slightest resistance as well as those who did not resist. In 1817, the British massacred the residents of Madulla for no better reason than the proximity of the village to a place where rebel Kandyans had attacked a British convoy with bows and arrows.[4] Twenty or more of the victims were shot by soldiers as they tried to escape from a cave. The treatment of some of the Sinhalese was, therefore, no different from the general experience of those targeted for civilization by European colonists. (The concept of an imported epidemic laying waste to vulnerable native inhabitants works just as well applied to the British, and other colonial nations, as it does to fungal spores.)

The hill country was perfect for coffee growing. At its peak, annual Ceylonese coffee production exceeded 50 million kg, which isn't much in today's terms,[5] but made fortunes for people with aristocratic names like Graeme Hepburn Dalrymple-Horn-Elphinstone. For a few decades, Ceylon was the world's top coffee producer, and the price for Ceylonese beans affected the rest of the coffee market. Exports to England began in 1802, but the first coffee estate wasn't established until the 1820s by Sir Edward Barnes, who was also the governor of Ceylon. Speculators invested millions of pounds in Ceylonese coffee, promoting a cycle of boom and bust in coffee prices that has shaped the market throughout its history. The investments financed the wholesale demolition of pristine hardwood forests in the highland districts of Dimbula, Dikoya, and Maskeliya. Giant teaks, ebony trees, elephants, flying foxes, and other species in this sweltering haven for biodiversity made way for plantation agriculture.[6]

William Knighton was a coffee planter and newspaper editor in Kandy the 1840s. He wrote that unbridled speculation in coffee encouraged deforestation and planting even at low elevations that produced second-rate harvests.[7] The resulting dilution of the crop with poorer beans spoiled the reputation of Ceylonese coffee in Britain, and the quality was further worsened by prolonged storage after roasting and grinding. Knighton's insistence that "the berry should pass at once from the roasting-pan to the mill, and thence to the coffee-pot," seems very familiar to those who roast and burr-grind their own coffee today in pursuit of that elusive perfect cup. There was another respect in which Knighton was ahead of his time. Most of us have vague notions about shameful behavior of Victorians in the colonies, but Knighton's first-hand account of planter society exposes the roots of today's stereotypes. Knighton recounted the story of a gentleman called Siggins who abducted the daughter of one of his workers ("I honored the rascal's daughter with a little attention"), flogged her father when he begged to reclaim her, and gave him "another dozen" when he had the impudence to talk of reporting him to the magistrate. Siggins was of the opinion that "the fear of the whip or the cane is the only thing that rouses them [his employees] to exertion." Another commentator, Lester Arnold, cautioned that the Ceylonese laborer was "hopelessly unambitious, except when famine spurs him," and noted that the workers were not attracted to labor in the jungle unless they were "kept in it by the magnetism of an Englishman's presence."[8] As I mentioned earlier, the British were proud that they "abolished" slavery in Ceylon.[9]

Years before his famous travels in the Blue Nile region of Africa with Florence von Sass, Sir Samuel Baker had been a coffee baron in Ceylon. He established a mountain resort in Nuwara Eliya, whose resemblance to a British village earned the name "Little England." He traveled throughout the country, reveling in the scenery and wildlife as he explained in his sensitively titled *The Rifle and the Hound in Ceylon* (1854).[10] In the less disturbing *Eight Years' Wanderings in Ceylon* (1855), he described coffee planting in the Kandyan hills, admonishing anyone who believed they could buy a few hundred acres of forest, plant coffee trees, and sit back with a gin and tonic and flyswatter while the "coolies" did the work.[11] Baker referred to the early years of coffee growing as a mania, with bankruptcies more common than fortunes as the difficulty of clearing the forest became apparent and the price of the commodity wobbled. But as the forests continued to be felled and torched, coffee continued to swell the bank accounts of many British investors in the second half of the nineteenth century. The aforementioned Graeme Hepburn Dalrymple, who later became Sir Graeme, was the largest coffee proprietor by the 1870s, running the Dalrymple estates in Kotmale. He was 28 years old when *Hemileia* arrived during the peak of bean exports;

by the time he was in his 40s, the coffee tree had practically disappeared from Ceylon.

George Thwaites, Director of the Botanical Gardens in Peradeniya, was the first scientist to become acquainted with the rust.[12] Coffee planters had shown him a few infected trees on an estate in Madulsima in May 1869, and by July, two or three acres of trees were dropping their leaves.[13] Thwaites inspected pale orange blotches on the underside of the leaves and noted that they had a powdery consistency and could be easily rubbed off. He recognized that he was dealing with a fungus and sent samples to his friend in England, Reverend Miles J. Berkeley.

Berkeley is one of a handful of scientific giants who escaped wider recognition due to their calling in mycology rather than sexier disciplines like astrophysics or paleontology. A curate with presumable devotion to Christian superstitions, he was nevertheless sensible in his study of plant diseases. Sweeping away farcical notions concerning the devil's influence during the potato famine, Miles was the gentleman responsible for isolating the microbial culprit for the Irish plague. I'll return to his role in the potato blight story in chapter 7, but for now, the story focuses on his serendipitous interest in Ceylonese fungi. Years before the appearance of the coffee rust, Berkeley had assembled a collection of more than 1,000 fungi from Ceylon from specimens mailed to his Northamptonshire vicarage. Berkeley found no matches to the fungus on the coffee leaves in his Ceylonese specimens, nor in his entire herbarium of many thousands of fungi, and he published a description of the new organism in *The Gardeners' Chronicle and Agricultural Gazette*.[14] The orange dust on the leaves consisted of masses of spores called uredospores that are characteristic of the rust fungi (fig. 3.2). Rusts are more closely related to mushroom-forming fungi than the ascospore-producing pathogens of chestnuts and elms. (Other rust diseases are discussed in chapter 6.) The name that Berkeley chose for the rust on coffee referred to the unusual shape of the spores—they are smooth on one side (*Hemileia*) and roughened on the other, looking like a clog fitted with running spikes (for a Dutch sprinter?)—and to its destructive effects (*vastatrix* means destroyer). At the time of Berkeley's publication, *Hemileia vastatrix* was just getting started, but once its fury against the coffee trees in Ceylon became apparent, the reverend's name for the beast seemed providential.

Understanding the role of the rust in damaging the coffee trees was complicated by the fact that the plant had so many other opponents in Ceylon.

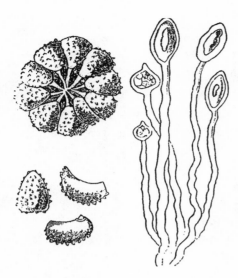

Fig. 3.2 The first illustration of the coffee rust fungus, *Hemileia vastatrix*. From M. J. Berkeley, *The Gardeners' Chronicle and Agricultural Gazette* November 6, 1869, p. 1157.

A pamphlet on these enemies published in 1880 detailed 6 bugs (or hemipterans), 14 species of moth and butterfly larvae, one ant, one fly, one locust, three beetles, one spider mite, and *Golunda ellioti*—also known as the coffee rat.[15] According to folklore, the rat attacked coffee trees every seventh year because this was when a particular species of plant blossomed and died, depriving the rodents of their normal staple. Whatever caused the rats to show up in the plantations every so often, their effects were undeniably bad for growers. The animals eschewed the berries but gnawed smaller branches back to the stem. On a more positive note, Malabar "coolies" were said to "fry the rats in cocoanut-oil, and convert them into curry," so some good came from the rodent's periodic arrival.[16]

The palm squirrel also deserves mention. The Victorian author of the pamphlet on coffee's enemies wrote that the squirrel "eats the berries, which, being indigestible with the exception of the outside pulp, are afterwards dropped and found upon logs and on the ground."[17] This is an intriguing reference in light of the fleeting interest in luwak coffee a couple of years ago. The luwak, or palm civet, is a nocturnal mammal that eats fruit, insects, smaller mammals, and, most importantly in Indonesia, coffee berries. The civet digests the outer pulp of the fruit, and defecates the beans unharmed, with the addition of a civety fragrance. Once roasted, the gamey brew from these beans supposedly justifies a price of more than $100 per pound, but I am concerned by the mathematics of this wonder crop. How many piles of civet feces must one pour through to find a pound of coffee beans? Web conversations suggesting that the entire story is a hoax seem more sensible than the idea of Indonesians praying for an outbreak of civet diarrhea. Perhaps you'll enjoy the following comments about luwak coffee on coffeegeek.com as much as I did: "think what a wonderful crappuccino you could make with it . . . It's the ideal accompaniment for a cup of 'coffee' that I received from a kiosk on Waterloo Station which resembled warm urine."

Returning to the colonies, the coexistence of all these coffee pests made it difficult to identify *Hemileia* as an unprecedented threat, and the earliest investigators mistook other fungi as the cause of disease. The hot, humid climate allowed numerous species to grow on plant surfaces, some forming fluffy webs over the coffee leaves, others growing in sooty mats fertilized by honeydew from the bountiful insects. Daniel Morris, Assistant Director at Peradeniya, initiated research on the coffee rust in Ceylon, but his investigations were cut short by a promotion and reassignment to Jamaica. His replacement was a 26-year-old scientist named Harry Marshall Ward, who was posted to the island in 1880 (fig. 3.3). By this time, the annual coffee

Fig. 3.3 Harry Marshall Ward. From F. W. Oliver, *Makers of British Botany. A Collection of Biographies by Living Botanists* (Cambridge, UK: Cambridge University Press, 1913).

harvest had plummeted, causing annual losses of around £2 million, or $2 billion in today's money. Ward had been invited to "attend the post-mortem" of Ceylon's coffee-growing industry.[18]

Ward was an unlikely academic success: a bright young man who made a life and reputation as a brilliant scientist despite leaving school at age 14.[19] While he shared little in common with the gentry who controlled the larger plantations, he felt tremendous empathy for many of the ruined growers who came from middle-class backgrounds. Ward soon solved how the fungus attacked the crop, but he concluded that the disease could not be beaten. He began his investigation by testing whether the spores described by Berkeley were responsible for infecting the coffee leaves. Working with the specimens shipped from Ceylon, Berkeley had reported that the fungus grew deep inside the leaves and produced its yellowish lesions on their underside. Coffee breathes through the lower surface of its leaves, which are perforated by stomatal pores, and *Hemileia* exploits these apertures. The spores germinate, extend slender tubes over the leaf, swell over the stomatal mouths, and ease themselves between the open lips (fig. 3.4). Ward discovered this by suspending drops of water on the underside of healthy leaves and carefully introducing a few of the spores into the drops on the tip of a needle.[20] He found that the characteristic lesions developed wherever he placed the spores and not elsewhere on the leaves. Moisture is critical for germination of the spores, and there was plenty of moisture during the summer monsoon season in Ceylon. But the rust spores are also tolerant of long periods without rain, which meant that the intermission of dry weather after the monsoon offered nothing but a brief respite for the coffee crop. The spores were rejuvenated by the first splashes of water and resumed work against the growers. Like most fungi, *Hemileia* is a champion spore-producer. Ward counted 60 or more "disease spots" per leaf, and each of these could shed more than 400,000 uredospores when the leaf was tapped by raindrops.[21] This means that 100 or so infected leaves on a single coffee tree could mist the air with a cloud of 2.4 billion infectious spores.[22] Even after the leaves fell from the sickened plants, they continued to shed rust spores. Ernest Large wrote that the spores "blew around the plantations numberless as grains of sand in the Sahara."[23]

After the rust has penetrated the stomata, it forms a branching colony that weaves its way through the junctions between leaf cells and reaches into the living cytoplasm with feeding structures called haustoria. A lot of

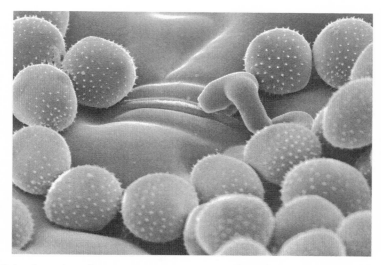

Fig. 3.4 Scanning electron microscope image of rust uredospores on a leaf surface. This species infects broad bean rather than coffee. One spore has germinated and its germ tube (or hypha) is attempting to penetrate a closed stomata. Photograph courtesy of R. Guggenheim and H. Deising.

fungal pathogens produce haustoria. Some are shaped like hands, others are just blobs, but they all form an intimate placentalike connection with the plant cell, indenting but not piercing the plasma membrane (fig. 3.5). In this manner, the fungus acts as a perfect parasite, absorbing its food from the hapless plant without killing it, at least not until it has taken all that the green cells can give. The attack seems subtle when compared with the swift vascular destruction wrought by the chestnut and elm pathogens, but the conclusion of the encounter between coffee and its rust parasite is equally deadly. Plant pathologists refer to this kind of fungus as a biotrophic parasite. *Hemileia* disables the coffee plant stealthily, not moving beyond its leaf tissue, just killing cells one by one, and finally causing premature leaf fall. A healthy tree keeps its leaves for 18–20 weeks; a rusted tree can lose them in 6 weeks. Without its solar panels the plant starves for lack of food.

To see how the pathogen was moving around in the plantations, Ward coated slips of glass with glycerine and clipped them to coffee branches. Rust spores stuck to these traps along with the spores of 50 other kinds of fungi. The moist air was thick with the fruits of fungal reproduction. Ward also discovered that the rust produced a pair of accessory spore types

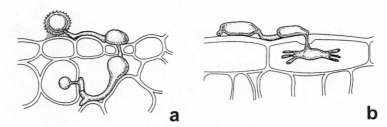

Fig. 3.5 Structures involved in plant penetration and feeding in (a) a rust fungus, and (b) a powdery mildew. (a) The rust forms an inflated structure called an appressorium above a stomatal opening before penetrating the leaf and feeding from living cells using its bulbous haustorium. (b) The mildew also forms an appressorium and then penetrates the cell wall of its plant host. After entry, the mildew absorbs the contents of the host cells via an elaborate haustorium equipped with fingerlike projections. Adapted from M. Hahn, H. Deising, C. Struck, K. Mendgen, in *Resistance of Crop Plants Against Fungi*, edited by H. Hartleb, R. Heitefuss, and H. H. Hoppe (Jena: Gustav Fisher, 1997), with permission.

besides the uredospores.[24] As the lesions on the leaves aged, they began producing a stalked cell with a thick wall known as a teliospore. When these teliospores germinated, they put out a short tube, divided it into four compartments with cross-walls called septa, and shed a third type of spore called a basidiospore from each compartment. Basidiospores are launched into the air by an intriguing mechanism called the surface tension catapult, which is also utilized by mushrooms.[25]

The production of the teliospores and basidiospores by *Hemileia* is a conundrum: nobody has been able to figure out the purpose of these spores in this species. Rusts in temperate zones use teliospores as survival capsules to endure winter temperatures and the temporary absence of juicy host leaves. In the coffee-growing tropics, winter doesn't present a problem and the pathogen can perpetuate itself indefinitely using nothing but uredospores. The basidiospores of other rust species infect a second, or alternate, host species. Ward was well aware of this fact as he worked to untangle the life cycle of *Hemileia*, but he couldn't find an alternate host— nor has anyone since.

There are a couple of explanations for the existence of teliospores and basidiospores in coffee rust. The first possibility is that an alternate host does exist, perhaps restricted to the birthplace of the fungus, but is superfluous when the organism finds a limitless supply of coffee plants. The other idea is that the production of teliospores and basidiospores is an evolutionary vestige, held over from the ancestors of *Hemileia vastatrix*, that no longer performs any useful function.

Ward's detailed reports on the rust offered a relentless stream of bad news for growers and for Ceylon. Differences in the severity of the disease between plants in a single plantation were related to their size. The photosynthetic capacity of a larger plant with lots of leaves sometimes enabled it to produce fruit after infection, but the rust eventually crippled every plant in the island nation. Growers hoped that certain coffee varieties would be resistant to the rust, but none was discovered. Citing the success of hop growers in Kent who had treated downy mildew, Berkeley, Morris, and others had advocated spraying the trees with sulfur-based fungicides. But unlike the mildew colony that grows on the surface of its host, *Hemileia* eluded the sprays by submerging itself in the flesh of the leaf.[26] Ward recognized that the fungus was vulnerable to spraying before it entered the leaves, and that this afforded only the briefest of opportunities for treating the disease immediately after the spores germinated. In any case, it was too late for this kind of preventive treatment by the time Ward surveyed the wasted plantations. Even if aggressive treatment of infected trees had been initiated when the disease first appeared in Ceylon, I doubt it would have been much use. Trees could be protected for a while through this kind of firefighting remedy, but the inferno was unstoppable: those uredospores would never have left the Ceylonese monoculture alone. Science offered understanding, but no solutions.

Alternatives to Ward's explanation of the disease were proposed along the lines of the rust being a sign of the plant's ill health rather than the cause of the disease symptoms: "Depend upon it the leaf disease is not *the* 'disease,' but an effect arising upon and from a *diseased* condition already contracted by the coffee trees. Fungus, blight, mouldiness, appear only upon already diseased subjects!"[27] This meant that the solution might be as simple as invigorating the plants with more manure.

Ward slapped these notions aside in his reports. His discovery of the cause of coffee rust resulted from the confident application of his scientific training. He clarified which type of spore caused the plant infection and proved a causal connection between the fungus that produced them and the symptoms of disease. Case closed, or so one might think.

In the 1890s, a bizarre idea, which became known as the "mycoplasm theory," was published by Swedish scientist Jakob Eriksson.[28] Rather than accepting that fungi invade plants from the outside, the new theory reversed the disease mechanism, allowing the fungus to appear from inside an already sickened plant. By now, you'll be familiar with the idea of a spore landing on a plant, or being implanted by an insect in the case of Dutch elm disease, then germinating and invading healthy tissues to injure the host. The whole field of plant pathology hinges on the following statement of fact: Bad stuff comes from outside the plant and makes it sick.[29] The mycoplasm theory was founded on a couple of mistaken observations. First, agricultural scientists had often been struck by the development of epidemic rust diseases in crops even after herculean efforts to remove infected plants. The reason that some of the rusts kept reappearing was that the fungi survived on one of those alternate plant species and found their way back to their original victims later on. This use of an alternate host was a well-known phenomenon that had been illuminated by the German scientist Anton de Bary in the 1860s. The reason for the persistence of other diseases, like coffee rust, was simply that there were no effective means to eradicate all of the infectious spores of the pathogen. Nevertheless, Eriksson pondered an alternative explanation: Might some diseases originate from inside an apparently healthy plant? He believed that he had made a groundbreaking observation that endorsed this idea.

Eriksson claimed to have identified fungi hiding within the cells of plants that expressed no symptoms of disease. He believed that these special bodies, which he named *corpuscules spéciaux* (*corpuscules spécieux*, or specious bodies, would have been closer to the truth) were the first signs of a surfacing disease. They served as Trojan horses, their presence signifying the imminent attack upon the plant. For a while, Eriksson surmised, the fungus persisted in some kind of harmonious symbiosis with the plant without causing it any harm, waiting for an opportune moment to get nasty. He was 100 percent wrong, and could have been responsible, had anyone of influence taken him seriously, of pushing plant pathology back a couple of millenia to the time when fungi were regarded as an "evil ferment of the earth."[30] Eriksson's "special bodies" were, in fact, cells of rust fungi that had developed from spores on the leaf surface. The spores germinated, sending out slender filamentous hyphae that penetrated the plant tissue, forming a colony that wormed its way through the junctures between the green cells. Every so often, branches of the fungus pushed their way into the interior of the plant's cells to begin sucking their juices with its haustoria. The corpuscules were nothing more than the haustoria of the rust caught in the early act of feeding.[31]

Eriksson's skills as a microscopist weren't stellar, and it seems that he consistently failed to trace the origin of the bodies in the interior of the plant back to the rest of the fungus that existed on its surface. He cut thin sections of infected leaves but failed to create a meaningful three-dimensional picture of the diseased tissue. When one of Eriksson's slices caught one of the rust's feeding cells, or haustoria, he didn't look carefully at the tissue slices from either side to see that it had been tethered to a larger colony. This might be difficult to picture, so imagine digging a trench into the ground and slicing your spade through a rattlesnake den. A sensible person would reason that he or she had disturbed a knot of angry reptiles that had slithered down a burrow to reach their subterranean winter quarters. According to Eriksson's illogic, however, one might wonder whether the snakes were some natural component of the soil, a *serpentent spéciaux*.

When Ward had seen Eriksson's drawings of his corpuscules spéciaux, he immediately recognized the type of feeding cells he had illustrated in *Hemileia*. He became the harshest critic of Eriksson, impaling the poor man in a brilliantly argued paper read before the Royal Society of London.[32] After his return to England, Ward was elected to the Royal Society, married, and enjoyed an illustrious career as the chair of botany at Cambridge. Unfortunately for the field of plant pathology, Ward died at the peak of his scientific creativity at the age of 52.

From the perspective of the ruined coffee growers, the scientific arguments about the rust were meaningless. Immunity from import taxation had given the Ceylonese growers a virtual monopoly in Britain, and Ceylon had been regarded as the greatest coffee-producing country for much of the nineteenth century. The microscopic fungus made a fiasco of this British economic miracle. In 1892, the heavily invested Oriental Bank collapsed. Faced with personal bankruptcy, expatriate families fell apart, growers fought among themselves, and some committed suicide. The plantations were left to rot, and "a failing culture with its cities vanished into the jungle."[33] In the face of the extinction of coffee in Ceylon, a few far-sighted growers recognized that the plantations could be turned over to tea growing. William Ukers wrote that "families ruined by coffee returned to Ceylon, took off their coats, and started with grim determination, which has been an example to British colonists ever since."[34] And while this wasn't a universal response to the catastrophe, Ceylon has, ever since, enjoyed the reputation as the exporter of some of the world's finest teas; having been weaned on gallons of the brew, I speak with tremendous authority.

Tea had been grown in Ceylon since the 1840s, but no more than 1,000 acres were planted until the blight's appearance. Between 1875 and 1895, tea plantations blossomed from 1,080 to 305,000 acres. Visiting the island in 1890, millionaire grocer Thomas Lipton realized that the rust-ruined land offered a terrific opportunity to grow inexpensive tea for sale in his British stores. He began by purchasing five ex-coffee plantations (later acquiring another dozen), and began growing tea. Rather than following the tradition of selling loose-leaf tea to customers from a chest, Lipton boxed small quantities for individual sale and pitched the product with the slogan, "Direct from the Tea Gardens to the Teapot." The venture was exceedingly successful, and Lipton has prospered for a century as the trademark for penny-wise flavorful tea.[35] Incidentally, Lipton, known as "Sir T" after his knighthood in 1898, acquired his interest in the retail business from his Irish parents. They had operated a small grocery store in Glasgow following their emigration from the potato famine in the 1840s. It might be said that Lipton was an improbable beneficiary of the ravages of two of history's worst fungal epidemics.

And now we come to one of those sweeping historical claims that has been reproduced in innumerable books but bears no relation to the truth. It has been said that by destroying coffee in Ceylon that the rust transformed the British into a nation of tea drinkers. The rust most certainly destroyed coffee, and the British are, without doubt, great tea drinkers. My grandfather served as a courageous example of a nation's commitment to a single beverage. Vernon Money was a decorated veteran of the British Expeditionary Force repulsed by the Germans in 1940, and, I am certain, is the nicest human I will ever meet. He downed 20 or more cups of tea a day, each with multiple spoons of sugar. But Vernon's enthusiasm wasn't rooted in a national change of beverage at the end of the nineteenth century, because the British had turned from coffee to tea in the eighteenth century.[36] The reasons for the switch are complex, but a lot of the problem with coffee lay in its expense and poor quality in Britain. The importers ignored William Knighton's nineteenth century plea for swift conveyance of the bean from roaster to the coffee pot, instead offering customers the powdered formula for a dreadful brew whose flavor paled in comparison with even a mediocre cup of tea. Besides the poor quality of imported coffee, tea owed much of its success in Britain to its convenience: it is so much easier to titrate the perfect concentration of tea than coffee. (My father has been working on the latter objective for decades, employing an astonishing collection of expensive brewing contraptions to—unfortunately—bring forth horrifying liquids.) So while coffee rust turned Ceylon into a country of tea growers, the effect of the epidemic on the British culture was, in reality, very limited.

ℰ

Hemiliea's nineteenth-century route into Ceylon is unknown, though it may have moved with the southwest monsoons from the Horn of Africa. Following its debut in Ceylon, the rust spread throughout mainland India, Java, Sumatra, and the Philippines, and early in the twentieth century it attacked plantations in East Africa.[37] The rust was following coffee which was following humans. The disease was spread by clouds of orange uredospores carried for hundreds of miles by cyclonic wind. This kind of long-distance dispersal, resulting in what plant pathologists call a single-step invasion, is a low-frequency event.[38] The vast majority of the quintillion spores wafting from the coffee plantations in Ceylon in the 1870s and 1880s never encountered an overseas victim. But the effectiveness of the process is inevitable whenever monocultures of susceptible crop are established. Coffee was introduced to Brazil in the eighteenth century. After the rusting of the Ceylonese plantations, Brazil became the world's biggest grower and has remained so ever since. There's little value in recounting the detailed history of coffee-growing in Brazil in this mycologically focused book, because the story is so similar to Ceylon's, but here's a brief summary: pristine wilderness went up in smoke (including the now critically threatened Atlantic rainforest), the services of more than one million slaves were engaged, and coffee barons made a great deal of money. *Homo sapiens*, once again, demonstrated its inherent insensitivity to other humans and the rest of biology on the plantations. But the agricultural history of Brazilian coffee didn't repeat the pattern established in Ceylon. Amazingly, quarantines imposed on imported plants barred *Hemileia* from Brazil, and until 1970 the coffee crop was rust free.

A young plant pathologist, Arnoldo Gómez Medeiros, discovered the disease in Bahia in 1970 when he "accidentally touched some rust-affected coffee leaves growing at the margin of a cacao plantation."[39] Though his brush with the spores may have been accidental, Medeiros immediately recognized what he had found because he had seen the disease during a trip to West Africa in 1967. Coffee rust had been reported in Angola in 1966. West Africa is a long way from Brazil, but the Angolan coast faces Brazil, and trade winds sweep in the appropriate direction to carry spores across the ocean. Traveling at a speed of 50–60 kilometers per hour, at an altitude of 3,000 meters, the uredospores probably hitched a ride across the Atlantic in a week or less.[40] Using a small airplane, Brazilian investigators trapped the uredospores at up to 1,000 meters above the plantations using the same kind of sticky slides that Ward had used to capture the rust at leaf-level in Ceylon. By the time Medeiros discovered the rust, it was scattered across half a million square kilometers of three Brazilian states. Coupled with evidence of widespread decay on individual coffee trees, the extent of the epidemic suggested that the rust had already settled in South America by the time he was looking at infected leaves on his African tour. Ironically, rust uredospores and Medeiros may have passed one another in 1967 flying in opposite directions high above the Atlantic.

The case for wind-transportation from Angola to Bahia is strong, but purely circumstantial. For this reason, it is possible that the rust was introduced to Brazil on infected seedlings. People can also carry viable spores on their clothing. If Gómez Medeiros had toured the rusted African plantations in the early 1960s, he might have been the unwitting courier himself. Plant pathologists can serve a excellent vectors of fungal diseases. One way or another, sooner or later, *Hemileia* was going to attack the Brazilian crop. But the fact that it didn't arrive until a century after the Ceylonese epidemic allowed the industry to survive in South America. By 1970, the coffee industry had an arsenal of fungicides at hand. After such a long journey, *Hemileia* must have been furious.

Sulfur- and copper-based compounds have been used since Ward's time to protect the leaves from the germinating spores, but more recent reliance has been placed on systemic chemicals that kill the fungus even after it has penetrated the leaves.[41] The systemic fungicide triadimefon, for example, targets the manufacture of sterols by *Hemileia*. Coffee rust, in common with other fungi, produces a fatty molecule called ergosterol that resides in its cell membranes. Ergosterol substitutes for cholesterol in fungal membranes, and with no arteries to clog, nor hearts to attack, the fungus is a very happy camper lipid-wise. By interfering with this fungus-specific biochemical pathway, the fungicide kills the rust but not much else, which is a good thing in terms of groundwater contamination. When too much triadimefon is sprayed on a coffee plant, however, it responds sometimes by dropping its leaves, so the dosage is critical. Fungicides, and the labor needed to apply them, are expensive. Up to five sprayings are needed every year to restrain the persistent rust, accounting for one-fifth of the total production costs.

Quarantines against the movement of plants that might carry the rust are a crucial defense, but the long-term effectiveness of these measures is questionable given that *Hemeleia* has now caused disease in every coffee-growing region except the Hawaiian Islands. Cultivated coffee has very limited genetic diversity, ranking it among the world's most vulnerable monocultures.[42] Arabica coffee, as I explained earlier, came from Africa and was cultivated in Yemen. The Dutch smuggled the plant out of Yemen in the seventeenth century (this was a covert operation because the Arabs didn't want to lose their monopoly) and grew it in Ceylon. The Portugese may have done the same thing even earlier when they controlled the country. The Dutch also took coffee to Java, and from Java back home to Amsterdam, where they grew it in the botanical garden. In 1713, the Dutch gifted the Sun King, Louis XIV, with a single coffee tree that he nurtured in a specially constructed greenhouse in Paris. Coffee is self-fertile, so an individual sets viable seeds after flowering. With the assistance of the British and the Spanish, seeds from that plant in Paris were eventually spread across the tropics and serve as the source of much of today's cultivated arabica coffee.[43] Two distinct varieties of arabica coffee are recognized by botanists: *Typica* is the one that came from France via Amsterdam, and the second, called *Bourbon*, was introduced from Yemen to the Bourbon Islands (now Réunion) early in the eighteenth century. Every pitiful coffee-addict whose first waking thought is, "Coffee . . . yes, I shall get out of bed," is dependent upon the inbred freak offspring of a handful of lonely coffee trees that accepted the hospitality of the French aristocracy in Paris and the deforested soil of Réunion. The entire crop is almost universally susceptible to the rust, which is a chilling fact. I remember learning in school that the French made "coffee" from acorns during the deprivations of the Second World War. As a child I could not imagine wanting coffee that much, but if the Nazi's showed up in my town today I would, if there were a soupçon of promise in the experiment, make espresso from cow dung.

Plant pathologists are working to develop a disease-resistant variety of arabica coffee. Coffee rust can attack other species of coffee besides *Coffea arabica* (it has as many as 100 to choose from), but the West African *Coffea canephora* and *Coffea liberica*, the sources of robusta and Liberian coffee, are far less susceptible.[44] Robusta coffee is the stuff that is mixed with arabica coffee to produce many of the ready-ground products sold in cans and jars, and it is sold alone as that "use only during caffeine emergency" stuff known as instant coffee. Robusta beans are also blended with arabica in many premium espresso brands, so there is no justification beyond stupidity for a snob like me to turn up his nose at the mere mention of the bean. Robusta may even hold they keys to protecting the arabica variety because its DNA contains an array of genes that confer resistance to the fungus. Researchers have been working with naturally occurring hybrids between arabica and robusta coffee for decades, with the hope of creating an arabica mutant with robusta genes that offers rust-free crops of delightful beans without the continual use of fungicides. Having said this, however, *Hemileia* will probably catch up after a few years, with its own mutations to overcome any firewall erected by a hybrid coffee. Rusts never sleep, but evolve continuously, and they patrol the agriculture of this windy planet 24/7, with their numberless drifting spores.

Along with his other accomplishments, the mycologist-hero of this chapter, Marshall Ward, is credited with unmasking the fact that plantation agriculture accelerates the spread of pathogens by crowding the feeding microbes with a superfluity of easily assailable victims. Unless they are checked by a change in the weather or by a crop-spraying aircraft, infectious spores can spread outward from a single plant like the ripples from a pebble thrown in a pond. In its original setting, arabica coffee would have been separated from its kin by other plant species, so its spores would have encountered plenty of physical obstacles before they alighted on their next meal. *Hemiliea* is believed to have lived in this way for millions of years, as a low-density disease on wild coffee trees in Ethiopia, and possibly in Sudan and Kenya too. Genetic variation between the coffee plants may have acted as an additional challenge to the rust, with some varieties showing greater vigor in the face of infection.

Contemporary growers who specialize in shade-grown coffee seek to replicate some of the original ecological setting and natural resilience of their crop plants, but any savings in fungicide application must be weighed against the cost of lower production.[45] The market for shade-grown coffee is increasing, but the minority of consumers willing to pay more for the promise of sustaining tropical biodiversity hasn't changed the practice of growing the crop in full sunlight in much of the world.[46] One strike against shade-growing, as far as growers are concerned, is that its efficacy against the rust remains unproven. Studies have demonstrated that another disease, brown eyespot, caused by the fungus *Cercospora coffeicola*, is actually encouraged by shade growing.[47] Investors in the vast full-sun plantations in South America are delighted to report that their crops haven't succumbed to a major outbreak of *Hemiliea* in many years. The reason for this isn't clear. Perhaps the widespread use of fungicides has knocked the fungus into submission; perhaps it will be back before long. Regardless of the future history of coffee-rust epidemics, our collective swallowing of a gobsmacking 3,300 cups of coffee per second forecasts a continuing addiction to monocultures of caffeinated beans.[48] It is getting late. Time for an espresso, I think.

CHAPTER 4

The Chocaholic Mushroom

My neural demand for caffeine is accompanied by an unremitting lust for chocolate. This addiction began as a sense of desperation as I waited for my father to hand over slivers of Mars bars when we watched television (he chilled the confection in the refrigerator to enable miserly slicing), was heightened when I earned enough money as a teen to purchase bags of bars that were consumed secretly in disgusting quantities, and has mellowed in adulthood so that it can be assuaged with a little of the pure stuff secreted in a kitchen drawer. By "pure stuff," I mean the 85–99 percent cacao bars made by companies like Lindt and Scharffen Berger. Scharffen Berger, based in Berkeley, California, takes the purity theme further with their delectable bags of cacao nibs.[1] Not for the squeamish, these fragments of roasted, shelled cacao beans taste more like nuts than chocolate, but they do convey a bright foretaste of their usual sweetened destiny.

Writing in the 1890s, John Hinchley Hart, superintendent of the Royal Botanic Gardens in Trinidad wrote, "Fortunately for the cultivator the serious diseases which at present attack the Cacao tree in the West Indies are few."[2] Unlike coffee—the grower might have fantasized—cacao has no competition from fungi: we can cultivate and consume cacao in carefree abundance. This must have pleased Daniel Morris. Remember him from the previous chapter? He was the early investigator of coffee rust who had been promoted to Jamaica. Morris ended up as the Imperial Commissioner of Agriculture for the West Indies and visited Hart in Trinidad to learn about cacao. Having escaped the rust-infested plantations of Ceylon to find an equally lucrative but unmolested crop must have been a big relief. But the respite was temporary. In 1911, Hart revisited the issue of diseases in *Cacao: A Manual on the Cultivation and Curing of Cacao*, which included a table of 18 fungal afflictions.[3] The first of these, *Phytophthora* pod rot, had been discovered in Grenada in 1895,[4] and others followed in swift succession, including *Diplodia* die-back, multiple root diseases, and witches' broom. Evidently, something unpleasant had happened.

In a reprise of the coffee story, the development and geographical diffusion of the cacao crop was followed by the arrival of fungal pests. There had been far earlier problems with cacao, beginning with a terrible epidemic, referred to as "a blight attacking the pods under certain atmospheric influences," that drove through the Spanish plantations in Trinidad in 1727.[5] Cacao had already been grown on the island for 200 years and had become a valuable export at the time of the first blight. We can't be sure of the disease agent, and some sources refer to crop damage by a devastating hurricane. But the description of pod blight is more consistent with the *Phytophthora* rot that I'll describe later. The crop was abandoned for 30 years until the Spanish reintroduced cacao to Trinidad by planting a different variety called "forastero." The original variety was "criollo," which was one of the gifts brought to the Old World from South America by Hernán Cortéz. Criollo originated in the Venezuelan foothills of the Andes but had been domesticated by the Aztecs long before Cortéz acquired its beans in exchange for his crew's gift of smallpox. Forastero was a hardier variety that evolved in the Amazon and had a greater degree of disease resistance than criollo.[6] A third variety called "trinitario" arose as a spontaneous hybrid (meaning no human intervention was involved in its birth) between the few surviving criollo plants and forastero. Trinitario was also hardier than criollo. By the time the British controlled the island, the epidemic was a distant memory and the new variety was thriving, so Hart can be forgiven for his sunny view of the commodity.

Cacao beans, or seeds, are produced in massive fleshy pods that dangle on short stalks that stick out from the stems and branches of an evergreen tree named *Theobroma cacao* (fig. 4.1). *Theobroma* means "food of the gods," reflecting the Mayan belief in the divine origin of the tree. Cacao is the correct spelling for the tree and the beans, but "cocoa" is easier to pronounce. Being the fruit of the cacao tree, the pods develop from the ovaries of the cacao flowers. Cacao flowers are very pretty. In his book *The Chocolate Tree*, Allen Young describes them as "bold white stars" that stand out "against the blackish branches."[7] The flowers are short-lived, opening by sunrise and falling off the tree by the end of the second day unless they are

Fig. 4.1 Leaves and fruit of cacao, *Theobroma cacao*. From A. Gallais, *Monographie du Cacao ou Manuel de L'Amateur de Chocolat* (Paris: Chez Debauve et Gallais, 1827).

pollinated. Insects are the natural pollinators, but the original relationships between certain rainforest midges and cacao flowers have been upset by growing the crop far from home in plantations. Fewer than 1 in 20 flowers bear fruit.[8] This seemingly low success rate is comparable, however, with the productivity of wild plants, which proves that other insects are effective substitutes for the native forest pollinators.

The minority of flowers that receive sperm-carrying pollen grains from the bristles of an insect generate 30 or more almond-sized seeds, or beans, in pods that turn bright yellow or red when mature. A big cacao pod has the size and shape of a football and has deep furrows that run from end to end. A factory in China makes footballs from orange plastic foam that look exactly like cacao pods. (You must have seen these, strewn across America, replete with bite marks, lying in messy yards where they give miserable dogs brief pleasure in their tethered lives.) The conversion of beans into chocolate is relatively simple, or at least it seems so to someone who has never tried to do so. You begin by hacking the pods from the trees using a machete. The same implement can be used to crack open the pods and release the pulp-embedded mass of seeds. The pods are discarded, and the seeds are set in a wooden box for fermentation. A soup of yeasts and bacteria digest the pulp during the next few days, cause the abortion of the embryos inside the seeds, reduce their bitterness, and enhance the chocolate flavor. After this brewing process is completed, the seeds are removed from their boxes and dried in the sun or in a customized dryer. Penultimately, the thin husks surrounding the seeds are removed, and the nibs are produced by grinding the fleshy stuff within. Finally, the nibs are combined with other ingredients to produce chocolate.

At least 80 percent of the cacao grown today is the forastero variety and 10–15 percent is trinitario, with the balance going to criollo. The quality of the beans follows an inverse relationship to the amount grown: criollo is the finest cacao variety, forastero lacks delicacy and can be unpleasantly bitter, and trinitario, reflecting its hybrid origin, occupies the middle ground. A useful comparison can be made between cacao and coffee varieties, with criollo corresponding to arabica and forastero corresponding to robusta in terms of flavor, cost, and disease resistance. The bitterness of forastero cacao and robusta coffee is due to the accumulation of chemicals called phenols that may have evolved to discourage insect pests in their native rainforests. The purported health benefits of phenolic compounds, as powerful antioxidants, translates to new marketing opportunities for chocolate manufacturers.

Much is made of the ecological friendliness of cacao plantations. The majority of the world's cacao is grown on small farms under big trees, or at least beneath trees big enough to shade the cacao. This practice is an example of agroforestry, and it supports far greater biological diversity than "technified cacao" plantations where nothing but cacao is grown under the tropical sun. The ecological benefits of agroforestry are pretty obvious. The richest species mixes are maintained when cacao is cultivated beneath the giant trees in a rainforest.[9] In these rustic plantations the degree of forest disturbance is variable. Small farms of widely spaced cacao trees cultivated by Native Indians in the Amazon Basin look like untouched rainforest to the uninitiated, whereas forest trees and shrubs in more populous parts of Brazil have been thinned to a tenth of their original abundance to make way for cacao. Farms in which shade trees are planted alongside cacao are also widespread, and studies show that these also sustain more animal species than the technified farms that are becoming more common in Malaysia,

Colombia, and Peru. Overall, the spread of cacao has been less destructive to the environment than other tropical crops, and probably accounts for no more than one percent of the rainforest lost to agriculture.[10] But like every other agricultural practice, cacao is still bad news for biodiversity. The richest species mix in a forest is always promoted by the absence of humans.

Cacao beans are a commodity of global importance with annual production exceeding 3 megatonnes, worth more than $2 billion.[11] Exports of all chocolate products are valued at $7 billion, and consumer spending on these products competes with the retail value of coffee. West Africa dominates cacao production: two-thirds of all cacao is grown in Côte d'Ivoire, Ghana, Nigeria, and Cameroon, and the entire crop is forastero.[12] The reason for the dominance of this variety is its resistance to fungi.

Here's a brief review of cacao cultivation following the Spanish Conquest. The Spanish were very effective at boosting cacao throughout Central America by employing forced labor. The economics of the practice were unbeatable: settle the land, enslave the natives, and compel them to grow cacao on their own land.[13] Production declined in the sixteenth century because the native laborers were debilitated by syphilis and dying from smallpox and other introduced plagues. Cultivation of cacao in South America began after the sixteenth century and soon spread throughout the Caribbean. Today's distribution of the crop overlaps the broad tropical coffee belt. Introductions of the cacao in Southeast Asia followed the crop's decline in Central America; cacao made its way from Indonesia to Ceylon at the close of the eighteenth century, and African production began in the nineteenth century.

Until the mid-1990s, Brazil was second only to Côte d'Ivoire in cacao production; then came a fungus called *Crinipellis perniciosa*, and Brazil dropped to fifth place, behind Côte d'Ivoire, Ghana, Indonesia, and Nigeria. *Crinipellis* attacks cacao trees, covers them with monstrous growths called brooms, and ruins the lives of growers. It is, in short, a very nasty fungus.

Witches' broom, or *escoba de bruja* in Spanish, refers to sproutings of swollen branches tipped with bunches of bedraggled and stunted leaves (fig. 4.2).[14] I think that the name is lousy, because one must drink heavily before recognizing resemblance between diseased cacao and brooms, but I'll come back to this shortly. These misshapen growths ("brooms") develop from the typically leaf-bearing branches and from the stem-borne flowers. *Crinipellis* also attacks the pods, destroying the internal tissues and

Fig. 4.2 Symptoms of witches' broom disease caused by the fungus *Crinipellis perniciosa*. Adapted from R. E. D. Baker and S. H. Crowdy, *Memoirs of the Imperial College of Tropical Agriculture* 7, 1–28 (1943), by Katy Levings.

killing the beans. The stalks that support the pods become thickened, and the entire fruit is distorted. This kind of damage goes beyond the direct effects of fungal decomposition. These *Crinipellis*-induced grotesqueries are probably due, in part, to the infections' disturbance of the hormonal balance within the cacao plant[15]—imagine massively swollen genitals in response to an infection of your pituitary gland and you'll grasp a human analogue of this plant disease. Early in the infection, the fungus produces bloated hyphae that snake their way into the tissues of the cacao plant, extracting nutrients and inducing the brooms, but without killing the cells. This is just like the initial stage of coffee rust and is referred to as the parasitic or *biotrophic* phase of the disease. After several weeks, the fungus changes its growth form, producing thinner, more destructive hyphae that rapidly degrade host tissues. This is called the *necrotrophic* phase, reflecting the reliance of the fungus upon dead or dying tissues. Think of a killer who toys with his victims before turning them into a stir-fry, and you'll be unlikely to forget this lesson in plant pathology.

When the brooms become dry and brittle, the fungus prepares for departure by emerging from the plant in the form of tiny, crimson-tinted, white mushrooms and sheds clouds of basidiospores in the cool night air (fig. 4.3).

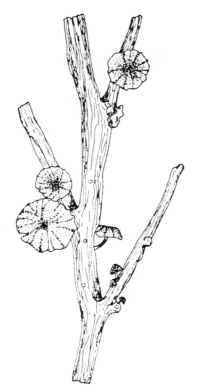

Fig. 4.3 Fruiting bodies of *Crinipellis perniciosa* sprouting from a cacao broom. These little mushrooms are the visible part of the fungus, whose feeding phase is buried within the tissues of the cacao plant. Reprinted with permission from D. N. Pegler, *Kew Bulletin* 32, 731–736 (1977).

These basidiospores are the only spore type produced by this fungus and are responsible for infecting the cacao trees. A single *Crinipellis* mushroom can release tens of millions of spores.[16] *Crinipellis* is one of few mushroom-forming fungi, or "agarics," that get much attention from plant pathologists. *Armillaria mellea*, the honey fungus, is the best known pathogenic mushroom. Colonies of this fungus, and closely related species, invade the roots of oaks and hundreds of other hardwood trees, shrubs, fruit crops, and conifers. But most of the agarics are involved in the decomposition of dead wood or partner with living trees in the formation of mutually supportive mycorrhizal associations.

The white button mushroom, *Agaricus bisporus*, sold fresh and canned in grocery stores, is a close enough relative to *Crinipellis* to bear comparison. Colonies of both species display spore-shedding gills beneath the caps of their mushrooms. But there are other kinds of wild mushroom that look a lot more like the witches' broom pathogen than those white buttons. At one time after its discovery, the cacao slayer was known as *Marasmius per-niciosus*, reflecting similarities between its fruiting bodies and those of the many species of *Marasmius* already described by mycologists.[17] Species of *Marasmius* are very common, but most are easily overlooked in the woods because they produce such tiny mushrooms. They grow on decaying leaves and well-rotted sticks. One that is often found growing from buried sticks on lawns is *Marasmius androsaceus*, whose common name is the horsehair parachute.[18] The "horsehair" part of the name refers to the black wiry stem, whose strength serves as a diagnostic feature, and "parachute" describes the white parachutelike cap that sits at the top of the horsehair. The common name offers a perfect snapshot of the organism. This fungus looks awfully similar to the little bells of the *Crinipellis* mushrooms scattered on the surface of dead cacao stems, which begins to explain why it was referred to as *Marasmius* for a while. One more thing about *Marasmius* mushrooms is that they are *marcescent*, which, for lovers of the impenetrable Saturday *New York Times* crossword, means to wither without falling off. In the context of mushrooms, the term refers to the ability to dry out, then resume normal activities when water is available. Some fungi are very good at this trick: drying to a crisp, but reviving with a splash of water and restarting spore release.

Like its benign relatives, *Crinipellis* can be found on the dead stems and twigs of other plants surrounding an infested cacao plantation. It flourishes on the woody tissues of plants that come in contact with the infected brooms, probably moving from cacao in search of more food. *Crinipellis* likes lianas, those often massive, vinelike flowering plants that clamber up tropical trees.[19] But how and why did *Crinipellis* evolve as a cacao pathogen?

Besides cacao, *Crinipellis* also attacks species of *Herrania*. These plants are members of the cacao family, the Stercauliaceae, and produce pods like those of cacao. *Herrania purpurea*, a resident of Central and South America, is known as monkey cacao because its fruits are eaten by monkeys.[20] The fact that *Crinipellis* assaults species besides cacao means that its behavior cannot be ascribed to the intensive cultivation of cacao. In other words, cacao isn't debilitated in some way by cultivation that exposes it to infection by normally benign forest mushrooms. *Crinipellis* is able to operate as a saprobe, feeding on a variety of dead woody tissues, but it obviously has particular talents as a parasite of the cacao family. This is an important distinction because there are normally harmless fungi that will colonize the tissues of plant and animal hosts weakened by some other factor. Such fungi are called opportunists, and their infections are opportunistic. *Crinipellis* doesn't fit this description. It is a truly pathogenic microbe.

Although plants lack the array of immune defenses available to animals, they do deploy a battery of defenses in response to the appearance of harmful fungi in their tissues. Plants respond to the presence of a fungus within minutes, generating hydrogen peroxide that inhibits the action of certain fungal enzymes, and sealing off the cells directly surrounding the invading hyphae. This reaction is known as the hypersensitive response and is an example of programmed cell death, or *apoptosis*, an act of cell suicide for the benefit of the larger organism. By killing a tiny proportion of its own cells, the plant may succeed in starving the fungus before it can spread. When *Crinipellis* is engaged in its biotrophic phase and the cacao brooms are still green, the accumulation of distinctive plant proteins signifies an ongoing battle at the molecular level.[21] But despite these defenses, *Crinipellis* is one of the fungi adapted to survive the hostile environment of the cacao's tissues; I discuss others in the parade of chocolate-loving fungi later in this chapter.

The earliest unequivocal report of witches' broom disease appeared in 1785. Alexandre Rodrigues Ferreira, a naturalist and explorer who mapped the Amazon Basin, described the symptoms of *Crinipellis* infection on cultivated cacao 500 kilometers up the Rio Negro from Manaus.[22] He described ugly growths on the plants as *lagartão*, or lizards, which is much more evocative of the actual disease symptoms than anything a witch uses for transportation. A century after Ferreira's expedition, the first widespread outbreak of the disease was reported in Surinam.[23] Plantations had been established by the Dutch in the eighteenth century, and exports peaked at 4,500 tonnes before *Crinipellis* was discovered in 1895. Within a decade of the mushroom's arrival, production crashed by 80 percent, fortunes were lost, and so on.[24] The harvest rebounded a bit in subsequent years, partly through improved drainage that produced more vigorous plants, but the disease never disappeared. Even with today's fungicides, Surinam remains a minor player in the cacao world. After Surinam, witches' broom appeared in British Guiana (Guyana) in 1906, where it eradicated the crop entirely; in Colombia in 1917; and then in top-producer Ecuador in 1921. The appearance of the mushroom must have been a particular shock to the Ecuadorian "cocoa kings" living north of the capital of Guayaquil, who enjoyed lifestyles more often associated with the exceedingly wealthy rubber barons of South America, whose fate I discuss in chapter 5. Gentlemen who had been sending their shirts to be laundered in Paris "went bankrupt instantly."[25] *Crinipellis* debilitates, but doesn't kill, the cacao tree, and the severity of the disease varied from year to year after its introduction to Ecuador. Since then, the cultivation of carefully selected hybrids of the forastero variety has allowed the country to rebuild cacao production in the presence of the fungus, and Ecuador ranks among the top ten exporters today.

Brazilian cultivation was centered in the coastal state of Bahia, far from the ancestral home of cacao and its mushroom in the Amazonian rainforest. In the decades after the Ecuadorian disaster the fungus continued to spread through Latin America and the Caribbean, but the plantations in Bahia remained free from disease. *Crinipellis* had, in all likelihood, lived with its host for millions of years in the Amazon Basin, and it hadn't amounted to more than an occasional irritant for the Indians who harvested cacao from the thinly distributed trees. Construction of the Trans-Amazonian Highway in the 1970s stimulated a phenomenal increase in Brazilian cacao cultivation.[26] Initially, plantations in the rainforested states were exceedingly productive. The varieties planted in this region were similar to those that had proven successful in revamping production in Ecuador, and Brazil was on course to become the world's biggest exporter of cacao beans. But with the proliferation of cacao occurring in the birthplace of the crop and its enemy *Crinipellis*, the planters' optimism was short-lived. Witches' broom became cacao's companion throughout the Amazonian plantations, and production fell sharply. Planters pruned and sprayed infected trees in an attempt to eradicate the fungus, but a survey of hundreds of farms in the Brazilian state of Rhondonia in 1982 found the disease on 90 percent of the plantations with older trees.[27] By the end of the decade, a third of the total Rhondonian crop was lost to *Crinipellis* every year. The disease was chronic and incurable.

There was, however, some good news for the Brazilian growers. While *Crinipellis* was feeding on cacao deep in the rainforests of the western states, its spores couldn't make the 2,600-kilometer journey to Bahia, which still produced 85 percent of the country's crop.[28] Unlike the spores produced by other destructive fungi discussed in the previous chapters, particularly those of the ocean-hopping coffee rust fungus, the basidiospores of *Crinipellis* are wimps. They dry out soon after discharge from their mushroom gills and cannot be resuscitated with a drink. Investigators who studied spore viability in the 1940s concluded that "one that does not cause an infection on the night that it is shed is probably killed the next day."[29] Harry Evans, a British expert on *Crinipellis*, estimated a maximum range of 100–150 kilometers for these spores, which was a favorable omen. The odds on containment were also enhanced by geography. The rainforest and the coastal plantations are separated by the vast stretch of spiny forest called "Caatinga," which offers a cacao- and pathogen-free buffer.[30] Clearly, the only way the disease could make its way to Bahia was with the assistance of a courier.

Quarantine measures were launched on major roads and at airports, including Manáus and Belem, linking Amazonia and Bahia.[31] With other diseases, caused by sturdier spores, someone might easily carry the infectious agents on their clothing. Witches' broom had to be carried on its host, which meant that the most obvious mode of transmission would be a truck bed filled with infected cacao saplings, or someone carrying infected beans. For a while, the quarantine held.

The first cases of witches' broom in Bahia were discovered on 112 trees on a single farm in May 1989.[32] The disease was focused on an area of a few hectares on either side of a major north-south highway (BR-101) close to the town of Uruçuca.[33] Someone had breached the quarantine. Efforts were made to contain the disease on the farm. Within three days of the disease outbreak, bulldozers were brought in to prepare a site for an operation center, and workers began to prune infected trees and burn the brooms. Knapsack mist-blowers and a helicopter were used to spray trees with fungicides. This monumental undertaking seemed successful: no further disease outbreaks were reported. The next step was to kill all of the cacao trees and vegetation in the central diseased area and in a surrounding buffer zone. The defoliant 2,4-D, also known as Agent Orange, was injected into trees, and brushed or squirted onto their debarked trunks. Finally, in August 1989, the management team decided to burn everything. The fate of the whole region seemed to hinge upon the containment efforts in Uruçuca.

In October 1989, plant pathologist João Pereira received a request from a cacao grower in the Camacan area, more than 100 kilometers south of Uruçuca, to look at infected trees. Pereira confirmed the grower's fears. *Crinipellis* seemed to have been at work on this farm for some time, migrating through a lush plantation within earshot of the trucks rumbling along Highway 101. The initial sites are also connected by waterways, raising the possibility that the dumping of infected cacao—pods or whole plants—in rivers could have pushed the disease to Camacan (or vice versa). Molecular biological analysis of *Crinipellis* from the two locations in Bahia suggests a different explanation. The technology of genetic fingerprinting shows that the fungi are different, which implies that infected cacao was brought to Uruçuca and Camacan on separate occasions.[34] Some commentators have raised the possibility that the pathogen's introduction was deliberate.[35] I doubt that this is true, though an investor might have made an astronomical sum of money by promoting cacao production in other countries with certainty of a crash in South America. My gut takes me to the more entropic explanation. Migrant workers from Rhondonia, or somewhere else in Amazonia, took infected plants to Bahia, and these were sold and planted in the two locations along Highway 101 in the mid-1980s. It is doubtful that anyone knew that the plants they were carrying harbored *Crinipellis*. The outbreak would have been forestalled, perhaps for decades, if Brazilians hadn't developed cacao plantations in the depths of the rainforest. But once cultivated cacao and its disease were widespread in the Amazon Basin, the migration of *Crinipellis* to Bahia was as inevitable as chestnut blight's airborne escape from the Bronx.

The result of the arrival of the fungus is unambiguous. Bahian plantations saw a 60 percent decrease in yield between 1990 and 1994, and national production fell from around 400,000 to 100,000 tonnes in the same time period. *Crinipellis* braked Brazil's rise among the world's top cacao producers, and the country slipped swiftly to its unenviable fifth place. Witches' broom exerted a tremendous sociological impact in Bahia. In the area surrounding the Port of Ilheus, the city that chocolate built, 200,000 workers in the cacao industry lost their jobs, which affected the lives of another 2 million people. Rural cacao-growing areas were swiftly depopulated, and the flood of bankrupted farmers into the cities led to a surge in homelessness and an increase in the crime rate.[36] The mushroom caused a national disaster.

The catastrophe in Brazil provoked unapologetic celebration in West Africa. Before witches' broom hit Bahia, Brazil and Ghana had jostled for the number two position among cacao producers behind Côte d'Ivoire. Today, with its annual crop approaching 500,000 tonnes, Ghana produces more than three times as much cacao as Brazil. Ghana, like other African producers, is *Crinipellis* free.

Cacao was introduced to Ghana by a former blacksmith named Tetteh Quarshie in the 1870s. Quarshie acquired the seeds during a trip to the Spanish possession called Fernando Po (now Bioko, which is part of Equatorial Guinea), an island 20 miles off the coast of Cameroon. Quarshie established a farm in the northern part of Ghana, and from this introduction, cacao soon replaced palm oil and rubber as the country's most lucrative cash crop. Ghana accounted for half of the world output of cacao and reigned number one producer for 60 years until it was surpassed by its neighbor Côte d'Ivoire in the late 1970s. Tetteh Quarshie remains a Ghanaian hero today, but no thanks to the son of Sir William Brandford Griffith, governor of the Gold Coast in the 1880s and 1890s. Griffith's son claimed that his father, rather than Quarshie, deserved the honor of introducing cacao to Ghana.[37] But in a pleasing example of British fair play, an official colonial inquiry in the 1920s ruled in favor of Tetteh's place in Ghanaian history.

Cacao farming in Ghana and Nigeria was built by African enterprise rather than by colonial edict. In Côte d'Ivoire, in contrast, French administrators pushed the crop upon reluctant farmers.[38] When Ghana was liberated from British rule (and from the name Gold Coast) in 1957, the ensuing unrest reflected in military coups and failed elections led to the neglect of plantations, and the cacao industry suffered decades of setbacks. Côte d'Ivoire's independence came in 1960, but cacao farming prospered under the control of the dictator Félix Houphouët-Boigny, who recognized the value of the crop. I said earlier that Ghana produces three times more cacao than Brazil, but Côte d'Ivoire has produced up to three times more than Ghana in recent years.[39] The relative success of Côte d'Ivoire and Ghana changes from year to year, according to political unrest in Côte d'Ivoire[40] and according to weather conditions and market factors in both nations. A great deal of cacao farming is done by poorly paid migrant labor, and the industry in Côte d'Ivoire is dependent on workers imported from Mali and Burkina Faso. Americans have little justification for acting blameless when discussing poorly paid migrant labor, but stories about child slave labor in African plantations have received widespread coverage in American newspapers. The flurry of interest in "chocolate slaves" led to calls for a boycott of Ivorian chocolate, but concern has been greatly dampened by investigations that dispute the prevalence of this practice. Mort Rosenblum offers a balanced appraisal of the issue in his book *Chocolate: A Bittersweet Saga of Dark and Light.*[41]

West Africa's dominance in cacao production is safe only as long as witches' broom disease stays in South America. The fragility of the *Crinipellis* spores, and the westerly flow of the trade winds, protect Africa from a Brazilian import, which means that human intervention offers the only obvious threat to the world's chocolate supply. Malaysia is another cacao-growing country that enjoys a *Crinipellis*-free industry. The threat to the world's supply of chocolate is, of course, real and pressing. As John Hedger, professor of tropical mycology at the University of Westminster in London, says: "All it needs is some asshole taking a tiny sample of witches' broom to West Africa or Malaysia. It will happen."[42]

The focus on witches' broom in this chapter is due in part to the unusual case of a mushroom-forming fungus causing a disease of global significance. There are, as I have mentioned, no other species from this otherwise benign group of fungi that cause so much damage. *Crinipellis* makes for a good story. But, in terms of total crop losses, a handful of other fungi also qualify for the "nasty fungus" appellation for their crimes against chocolate. Before discussing these diseases I must mention rodents. Like the rats that gnaw their way through coffee plantations, cacao is assaulted by a number of mammalian pests. Among a multitude of cacao-molesting rodents, chocolate-loving rats are startlingly diverse. A reference book devoted entirely to diseases of cacao lists the following pod-damaging species: the giant Gambian rat, Defua rat, spiny rat, gray-pouched rat, rusty-bellied rat, multi-mammate rat, water rat, rice rat, soft-furred rat, Tullberg's soft-furred rat, Pacific rat, Müller's rat, Norway rat, black rat, long-tailed cocoa rat, target rat, thicket rat, and the cane rat.[43]

Compared with rat pests, the number of fungal pathogens seems remarkably restrained. I'm going to discuss two more of them besides witches' broom. The first is called black pod disease, whose symptoms (not surprisingly) include the blackening of the pods. Worries about introduction of the microbes that cause black pod are unnecessary because they live everywhere that cacao is being grown. Worries about the disease itself are relevant, however, because black pod causes greater crop losses than witches' broom.[44] Black pod disease is caused by a handful of different species of *Phytophthora*, related to the pathogen that destroyed the Irish potato crop in the 1840s (potato blight takes center stage in chapter 7). Phytophthoras are water molds, more closely related in evolutionary terms to giant kelps and glass-hulled diatoms than they are to *Crinipellis* and the other fungi discussed so far. But because *Phytophthora* and other water molds behave like fungi, they have been studied by mycologists since mycology became a distinctive branch of science in the nineteenth century. Science writers have a great deal of difficulty in describing these microbes and resort to hideous combinations such as "funguslike protoctist," "stramenopile (or straminipile) fungus," and "oomycete water mold." In this book, I've opted for water mold, and refer you to *Mr. Bloomfield's Orchard* for more information.[45]

Compared with a rust like *Hemiliea vastatrix* that lives on nothing except coffee plants, the water molds that cause black pod aren't picky eaters. The most prevalent cause of the disease is *Phytophthora palmivora*, which attacks hundreds of plant species besides cacao. One of its alternative food sources is banana. Even if you dislike bananas, this is an unfortunate characteristic, because bananas are grown on African plantations interspersed with cacao. The water mold can be splashed from infected flowers and fruits of one plant to the other.[46] Ants and other insects also serve as black-pod vectors as they travel between plants.

The black-pod Phytophthoras can infect cacao pods at all stages of development and also feed upon the roots, stems, and leaves. In common with *Crinipellis*, the filamentous hyphae of the pathogen colony spread throughout the plant tissues. But the rapidity with which pod rot develops suggests that the Phytophthoras dispense swiftly with the fancy biotrophic phase of development in favor of killing and digesting every cell they encounter. After a satisfying meal, the water mold emerges at the surface of the pods and produces masses of microscopic oval-shaped structures called sporangia. As their name suggests, sporangia produce spores; each *Phytophthora* sporangium generates a clutch of swimming spores called zoospores. It's curtains for a healthy plant once it is splashed with these sporangia. Providing they don't dry out, the sporangia open and expel the swimming spores. The zoospores travel a few millimeters over the plant surface searching for a suitable entry point. Then they settle down, withdraw or jettison their swimming apparatus (a pair of flagella), and germinate by putting out a slender hypha that pushes into the cacao surface.

Cacao farmers have some hope of controlling black pod disease by some surprisingly simple but backbreaking efforts. The concentration of sporangia in a plantation can be reduced by frequently harvesting pods to limit the number that become colonized by the pathogen. Destroying infected pods and diligently pruning trees are two other effective husbandry practices. Spraying foliage with copper fungicides and chemicals such as metalaxyl that migrate throughout the plant can be very effective at treating black pod, but the cost of these chemical measures is prohibitive for many growers.[47] The severity of the disease can also be reduced by thinning the canopy of the plantation. The effectiveness of this method is due to the increased air circulation around the cacao plants and reduction in humidity: black pod disease creates the most problems for farmers in wettest parts of the topics during the wettest parts of the year. One might argue, therefore, that the cluttered agroforestry plantations are good for biodiversity but bad for cacao. This may be true when we consider black pod, but shade growing does have the advantage of limiting witches' broom.[48]

Research on cacao diseases focuses on the development of pathogen-resistant varieties of *Theobroma*, but there have been few breakthroughs to benefit growers in recent years.[49] One promising line of investigation concerns the use of fungi to combat fungal infections. This may seem strange—like treating a headache by hitting your head with a hammer—but there are fungi, called mycoparasites, that feed on other fungi rather than on plants. Mycoparasites that attack *Crinipellis* may someday prove exceedingly valuable agricultural products.[50] One such mycoparasite is *Trichoderma stromaticum*, whose aggressiveness against the colonies and mushrooms of witches' broom is very promising.[51] A somewhat different approach to disease control involves manipulating fungi that are normal inhabitants of plant tissues. These lodgers are called endophytes—fungi that live inside plants without causing any obvious harm to their hosts. The community of endophytes inside cacao is very diverse, and experiments have shown that these fungi can offer significant protection against black pod.[52] The hope, in this case, is that crop losses can be reduced by modulating the mix of endophytes.

Having written an uncharacteristically optimistic paragraph, I'll return to bleaker material with monilia frosty pod (sounds like an ice cream desert), the third and last fungal disease for this chapter. This is a badly misnamed affliction, because *Monilia* is the name of a fungal genus, and it doesn't cause frosty pod. By way of illustration, one might mistakenly assume that whooping cough is an illness contracted through intimate contact with an endangered species of crane. Whooping cough is, as you may know, caused by a bacterium, and monilia frosty pod is caused by a species of *Crinipellis* named *Crinipellis roreri*. For many years, plant pathologists thought that frosty pod was due to a type of *Monilia*, but research now shows that a strange relative of the witches' broom pathogen is responsible. This is a bit esoteric, but *Crinipellis roreri* is a kind of handicapped witches' broom fungus. I say handicapped because it doesn't produce any mushrooms, but instead subsists by generating the types of cells that produce spores in mushrooms (basidia) and using these cells as spores.[53] It is almost as if the frosty pod fungus gives birth to gonads rather than to fully formed offspring.[54]

Although *Crinipellis roreri* seems like one slice of a complete fungal life cycle, the fungus is highly effective as a cacao pathogen. Its colonies of filamentous cells grow between the cells within the cacao pod without triggering any obvious defensive reactions.[55] This "cryptic growth" can persist for months as the fungus dismantles the pod from within. Even when the pods are harvested, they often show no symptoms, which means that the fungus is spread unwittingly when infected pods are transported from the diseased plantation. When the fungus reaches the surface of the pod, it begins releasing its "spores," transforming the shiny fruit into a frosty bloom called a pseudostroma. The lack of a mushroom does nothing to limit its fecundity: an astonishing 44 million spores are shed from every square centimeter of the cream-colored pseudostromas. The disease has been around for a while. It aided witches' broom in the virtual extinction of cacao in Ecuador in the 1920s. Since then, frosty pod has invaded other South and Central American countries, and its arrival in Mexico is anticipated shortly. Globally, the fungus accounts for four percent of cacao losses, but it has devastated plantations in some regions. In Peru, for example, crop losses due to frosty pod are estimated between 40 and 50 percent. A knock-on effect of these losses in South America is the stimulation of coca growing (one letter makes all the difference), which reminds me to recommend the DVD collection of *Miami Vice*—the 1980s television program that stimulated my move to the United States in an utterly fruitless search for suitcases full of money and the keys to a Ferrari Testerosa. Twenty years later, I have a large mortgage and drive a Mini Cooper. But I digress, but not too far, because the next chapter concerns tires.

CHAPTER 5

Rubber Eraser

I was eyeing packages of prophylactics in the grocery store the other evening. As I stood in the checkout line, the product that caught my eye was labeled Magnum, which made me ponder the size of the items in the box. This reverie turned to alarm when I noticed packages placed a little higher on the shelf: Magnum XLs—extra-large Magnums. The reason for my dismay was mycological in nature.

The natural latex of the Magnum XL is an alarmingly vulnerable commodity. As a cherubic lad, I used to impress my friends by raiding coins from a condom dispenser in a pub toilet close to the hut where we met for Cub Scout meetings. A hard rap in the appropriate spot would usually release a little pocket money. I mention this childhood memory because I recall that some wag had scrawled "Beware of remoulds" (British spelling) on the machine. It was years before I knew what that meant. But here's the rub. If a fungus called *Microcyclus ulei* is introduced to Southeast Asia, Magnum XL users may be forced to recycle.

Rubber is serious business. Worldwide production of the commodity exceeds 20 million tonnes. This splits 60/40 in favor of synthetic rubber, which is a petroleum product, leaving almost 9 million tonnes of natural stuff that runs from the trunks of wounded trees in Thailand, Indonesia, Malaysia, and other nations in the tropical "rubber belt." Thailand is the biggest single producer of natual rubber, with annual exports approaching 3 million tonnes.[1] Even with the extra material needed for Magnum XLs, a far greater proportion of the natural rubber harvest is used to make tires. As it happens, I'm writing on my laptop in the customer lounge of car dealership this rainy morning, waiting, ironically, for a new set of tires. Sharpened by my evident dependence on rubber, my interest in *Microcyclus* is more than academic.

But I'll begin with a brief history of the commodity rather than its enemy. Rubber began, like cacao, as a tropical product appreciated by the inhabitants of rainforests. Rubber is derived from white, gooey exudate that flows from a special network of vessels that run underneath the bark of various tropical plants. The source of commercial rubber is a tall tree called *Hevea brasiliensis* (fig. 5.1). Native people sliced through the bark of this species, and other latex-producing trees scattered throughout their forests, and used the condensed exudate to fashion water containers. Columbus returned to Europe with rubber balls fashioned from the latex of another plant species in Haiti. In the eighteenth century, the astronomer Charles Marie de la Condamine described the traditional methods of rubber production in South America. (I would love to tell you that the word condom is derived from this Frenchman, but the etymology of the term is mysterious.) The chemist Joseph Priestly (1733–1804) recognized that rubber could be used to erase pencil marks, and this application explains the continuing British use of "rubber" for eraser. In case you're interested, the word for rubber in every other Indo-European language is derived from a sensible Amerindian name, *cachuchu*, that means "weeping wood."

Our contemporary reliance on rubber is rooted in the invention of vulcanization by Charles Goodyear. Rubber had been used to make life preservers, coats, shoes, and other products in the early nineteenth century. For the most part, these were troublesome wares because the rubber melted in warm weather and stuck to everything, became hard in winter, and tended to decompose and smell. Goodyear, with his associate Thomas Hayward, fixed this problem by heating rubber mixed with sulfur to transform the uncontrollable raw substance into a highly resilient and elastic material. The method was hard won, the award of a patent in 1844 marking the end of a desperate, decade-long quest by Goodyear. This stubborn optimist pursued his goal despite continual bankruptcy, the suffering of his wife and children (four of whom died), and his own debilitation due to lead poisoning. Goodyear was a fascinating gentleman, and his story is wonderfully told by Harold Evans and co-authors in their book *They Made America*.[2]

Interest in Brazilian rubber, known as Pará rubber for the state of Paraíba where it was discovered, grew swiftly after Goodyear's discovery. But it wasn't until 1865 that botanical studies of herbarium specimens of *Hevea brasiliensis* at the Royal Botanic Gardens, Kew, established the identity of the source of the imported latex.[3] At this time, during the Industrial Revolution, manu-

Fig. 5.1   Leaves, fruit, and seed of the rubber tree, *Hevea brasiliensis*, illustrated by Henry Wickham. From H. A. Wickham, *On the Plantation, Cultivation, and Curing of Parà Indian Rubber (Hevea brasiliensis): With an Account of Its Introduction from the West to the Eastern Tropics* (London: Kegan Paul, Trench, Trübner, 1908).

facturers were exploring new applications for the remarkable substance. Rubber served as a perfect material for pressure-tight gaskets in steam engines, and big lumps of the stuff were fashioned into buffers between railway carriages. The wild plant grew south of the Amazon River, spreading across a vast territory through Brazil to northern Bolivia and eastern Peru, where two or three large trees grew per hectare in the rain-soaked jungle.

Having mastered the rudiments of cacao harvesting in the previous chapter, I think we can tap a rubber tree in a virtual sense. As already mentioned, the latex courses just beneath the bark of the *Hevea* tree, so I'll chop at the trunk of mature tree with a hatchet and catch the latex in a bucket. This seems to be effective; I'll soon have enough rubber to make a pack of condoms for a chihuahua. If you want a set a tires, you'll need to employ someone with experience. To harvest more rubber from a tree, it is important to avoid severing the cambium layer beneath the latex-conducting channels, because this tissue is responsible for replenishing the phloem tissue damaged by tapping.[4] The modern tapper cuts steep, spiral-shaped troughs (top left to bottom right direction), running half-way around the trunk. When the bark is cut, the latex flows for a couple of hours before the tree seals its injury. The next day, the tapper reopens the channels by drawing a notched knife directly below the previous incision. When one patch of bark loses its vitality, the tappers begin to work another area of the trunk to allow the tree time to repair the damage. This practice of controlled wounding, referred to as the excision method, dates to tapping experiments in the 1890s by Henry Ridley, known to his colleagues as "Rubber Ridley" and, less affectionately, as "Mad Ridley."[5] Ridley served the British Empire as director of the Botanic Gardens in Singapore, and he earned his nicknames for his passionate promotion of rubber growing at a time when there was far greater interest in coffee.

The milky latex that runs from the tapped tree is half water, 10 percent resins, proteins, and sugars, and 30–40 percent rubber.[6] Careful tapping can yield 3 or 4 kilograms of rubber from a plantation tree every year, and a healthy plant can be drained in this fashion for 30 years. When you consider that at least 2 kilograms of natural rubber are needed for the manufacture of a single automobile tire, it is obvious why so many trees are needed.

The history of the rubber cultivation is far more interesting than its practice. With demand for rubber increasing in the 1860s, Clements Robert Markham, an official in Britain's India Office, conceived a plan to obtain seeds of the rubber tree from Brazil and to cultivate the plant in the British colonies in Asia. A few years earlier he had participated in the transplantation of quinine-producing cinchona plants from their native Peru to Ceylon and India. The price of the antimalarial drug fell sharply under cultivation, and its availability accelerated the European colonization of Africa and Asia. Markham wanted to repeat this success. Henry Alexander Wickham was the young hero, or rogue according to some, who executed Markham's plan by conveying thousands of rubber seeds out of the jungle and transporting them to Kew.

Wickham's expatriate life had been a disappointing experience. He dealt in bird plumage in Central America for a few years, harvested rubber from wild trees in the Orinoco basin, and established a farm in Santarém in Brazil. None of these enterprises were successful, but he unwittingly secured a brighter future by sending botanical specimens to Sir Joseph Hooker, the famous director of Kew Gardens, and also by writing about his South American adventures in his 1872 book *Rough Notes of a Journey Through the Wilderness*.[7]

References to rubber in Wickham's book caught the attention of Hooker and Markham: here was a gentleman, most importantly an English gentleman, who possessed an intimate knowledge of rubber trees and tapping methods, and, presumably, knew where he could collect lots of seeds. After a lengthy correspondence, Wickham negotiated a paltry fee of £10 per 1,000 seeds, and the bounty was gathered in 1876. Rubber seeds are flung from the exploding capsules of *Hevea* trees and must be collected from the forest floor before they are consumed by animals. This meant that Wickham had to act quickly. With the help of Tapüyo Indians whom he employed in a series of canoe trips up the Amazon (referred to as "rather ticklish work" by Wickham[8]), he collected more than 70,000 seeds, layered them between banana leaves in baskets, and loaded the precious cargo on a ship bound for Liverpool.[9]

The accuracy of Wickham's account of the seed-collecting expeditions and his escape from Brazil has been questioned by many authors.[10] Wickham wrote that he gave orders to the ship's captain "to keep up steam," while he obtained the necessary approval from Brazilian authorities to leave the country with "exceedingly delicate botanical specimens specially designated for delivery to Her Britannic Majesty's own Royal Gardens of Kew."[11] The implication of his order to the captain was that they might be forced to flee if their passage was denied. It has been suggested that Wickham dramatized the events, noting that the export of seeds via the Port of Belém would have been a formality in 1870s. But I'm going to side with this part of Wickham's story. He was in a unique position to understand the value of his cargo, both to the British Empire and, more important, to himself. Having invested tremendous effort in collecting the seeds and packing them for travel, he knew that unless they crossed the Atlantic swiftly, the whole enterprise would have been wasted and he would resume his anonymous life as a bankrupt planter. Formality or not, I bet Henry Wickham's heart was racing in that muggy office in Belém as he waited for the official to rubber-stamp his papers. His life hinged on this moment in 1876, and he knew it. Warren Dean describes the baskets filled with *Hevea* seeds as "the most fateful cargo ever to descend that river."[12]

If Wickham had lost his cool, what would the twentieth century have been like without tires and latex condoms? In his later life, Wickham embellished the story of his Brazilian exploits, saying that the boat had been loaded by stealth close to a gunboat "which would have blown us out of the water had her commander suspected what we were doing."[13] Photographs of him as an old man, after his knighthood for his service to the England, show the "planter hero" with an extraordinarily furrowed face

Fig. 5.2 Sir Henry Wickham. From J. Loadman, *Tears of the Tree. The Story of Rubber—A Modern Marvel* (Oxford: Oxford University Press, 2005), with permission.

and gorgeous snow-white mustache (fig. 5.2). I would love to have heard him tell his stories, as we swirled the brandy in our snifters at the club and puffed on fat cigars.[14] I hereby urge Merchant Ivory Productions to make a movie about the life of Henry Alexander Wickham.

The British Isles do not furnish the ideal climate for cultivating tropical plants, but with much of the globe belonging to Queen Victoria, there was no shortage of "British" sites for rubber planting in Wickham's time. The passage from England to the tropics presented more of a challenge. The transportation of plants by sea had always been a serious problem. One of the crew's complaints against Captain Bligh was that he lavished water on cargo of breadfruit plants at the expense of their ration. Plants perished even during brief voyages. But the timing of rubber's export was fortuitous because of a fantastic invention in the 1830s. Nathaniel Bagshaw Ward (1791–1868) was a London physician and amateur botanist whose love of ferns led him to perfect the archetypal terrarium, called the "Wardian case." The utility of the Wardian case for protecting plants during sea voyages was proven by experimental shipments of ferns and grasses to Australia. All of the plants survived the six-month voyage. Hooker was one of the first botanists to employ the cases, and the plant hunter Robert Fortune employed

Ward's invention to transport 20,000 tea plants from Shanghai to Assam in the 1840s. The concept of a miniature greenhouse for keeping plants hydrated was ridiculously simple, but revolutionary. On dry land, ornate versions of the cases filled with luxuriant ferns and orchids caused a sensation in home decoration, and adorned every middle- and upper-class home. The inventor also recommended Wardian cases for the improvement of the poor: nature in miniature would, he hoped, enable the unfortunate to appreciate the beauty of God's creation in the absence of photosynthesis in their dark satanic towns. Ward may have imagined a downside, however, when he hinted that extra light and heat coaxes *Cannabis sativa* to produce "secretions of a powerful and dangerous character."[15]

The initial plan for rubber planting called for the shipment of the Kew seedlings to Burma (Myanmar), but this was abandoned in favor of Ceylon. In August 1876, most of the plants were loaded into Wardian cases, taken by barge down the Thames, transferred to British India liners, and shipped to the Botanical Gardens in Peradeniya, Ceylon. Ninety percent of the plants survived their 5-week voyage and were received by the garden director, George Thwaites. Thwaites had been first to alert the botanical world to the threat of coffee rust, which had scourged this commodity from the plantations by the time the rubber seedlings arrived. As the rust pushed *Coffea arabica* toward extinction in Southeast Asia, Brazilian growers seized control of the global coffee market, just as, paradoxically, rubber trees taken from its rainforest were flourishing in Asia. By 1882, the plants in Ceylon began producing their own seeds, and these were exported to India, Burma, and Singapore. The sale of the seeds became so profitable, that planters didn't bother tapping their trees in Ceylon. Plants grown from Wickham's seeds at Kew had also been shipped to Malaya (now Malaysia) which began commercial sales of rubber in 1898.[16] Other European nations acquired seeds from Brazil, establishing rubber plantations in Java and a number of African countries, and Americans exported seeds to the Philippines.[17]

Meanwhile, tapping of wild trees continued in the Amazon, and surprisingly little effort was made by the Brazilians to turn to plantation production of *Hevea* to compete with the Europeans. The tapping business involved considerable exploitation of native labor—"the most criminal labor organization . . . by the most revolting egotism"[18]—which saved on wages, and the supply of wild rubber trees was, in practical terms, limitless, given the expanse of *Hevea* territory in the rainforest. There were no incentives for private businesses to invest in establishing rubber plantations when there was so much of the stuff oozing from trees in hundreds of thousands of square kilometers of jungle west of Belem.

The lifestyles of the rich and infamous "rubber barons" who controlled the commodity were without peer. In the shadow of the forest, 1,600 kilometers up the Amazon, a little-known outpost called Manaus had became the capitol of the industry. Taxes from rubber exports financed the transformation of the city. By the close of the nineteenth century Manaus had a state-of-the-art streetcar system (donated by American investors), a municipal gas and water supply, electric street lighting, all-night bars and restaurants, a racecourse, and a bullring. A spectacular $2 million Italian Renaissance-style opera house opened in 1897. The barons built ornate homes, used banknotes to light their cigars (once in a while, I assume, for the cameras), and, like the Ecuadorian cacao kings of the previous chapter, sent their laundry to Europe. I like the story of one millionaire's response to an ice and fuel shortage: "To chill two bottles of champagne for his dinner, he rigged up a kerosene-powered refrigerator for £40, then used £9 worth of absinthe to fuel his launch for a half-hour jaunt up the Yavarí River."[19] The author, Victor von Hagen, wrote, "Manaus had actually become El Dorado. Gold flowed like water through its streets. The whole city throbbed to the . . . dream of wealth."[20] Thousands of fortune-seekers, many of them from Europe, arrived in Manaus every week.

But Wickham's seeds would destroy this monument to capitalism. Malayan rubber hit the market and the Manausian economy went belly-up in 1912. Europeans crammed themselves into steamers and fled in droves, those facing bankruptcy paying for passage with their jewelry. Millions of people living in the Amazon cities and towns were "stranded along the river banks," daughters of ruined families were forced into prostitution, and some of the barons committed suicide; the larger towns were depopulated in a matter of days, and smaller ones were "engulfed by the jungle."[21] The rubber barons called Wickham "the Executioner of the Amazonas" for ending their monopoly, and contemporary web sites refer to his act as an illustration of "biopiracy."[22] The fault with their viewpoint, however, is that the majority of Brazilian agriculture is based upon plants abducted from other parts of the world.[23]

Rubber cultivation requires considerable patience on the part of the investor. *Hevea* trees are untappable for the first five years of their life, and are rarely productive until their seventh year. Noting these difficulties in the 1900s, however, a European estate manager, W. F. C. Asimont, trumpeted the potential for fortune: "The prices for plantation rubber are so stable, the probability of the supply exceeding the demand so very remote, and the possibilities of a satisfactory artificial substitute for rubber being invented so improbable, that we are quite justified in assuming that the cultivation of Hevea Braziliensis for the next fifteen years, at least, will be one of the most stable and profitable ventures in which capital could be invested."[24] Asimont was writing from his experiences in Malaya, but his message wasn't lost on everyone in Brazil. Rubber was cultivated on a small-scale in Brazil before and after the glory years in Manaus, and a plantation industry would certainly have evolved in the Amazon basin in the twentieth century if a certain fungus hadn't materialized in nearby Suriname (then Dutch Guiana).

Coastal Suriname is 900 kilometers from the place where Wickham collected *Hevea* seeds, but the first rubber seeds planted there in 1897 came from much farther away: Ceylon by way of London. Only nine trees resulted from this import, but traffic in seeds between Suriname, the neighboring colony of British Guiana (now Guyana), and Brazil spread the crop during the following years. Rubber farming flourished among cacao planters in Suriname who had lost their fortunes to witches' broom caused by *Crinipellis*. It didn't last long. Early in the twentieth century, botanists described symptoms of severe leaf disease in *Hevea*. Nursery plants in the botanical garden at Belém were "seriously injured," seedlings were defoliated in Suriname, and there were reports of disease outbreaks in nurseries and plantations elsewhere in South America. Discussing these early disease outbreaks in 1914, Thomas Petch, Mycologist to the Government of Ceylon, closed a prescient article with the following gloomy comment: "It is unnecessary to point out the bearing of this on the proposal to establish plantations of Hevea in its native country."[25] Planters ignored the problem for as long as they could, ascribing the symptoms of leaf disease to the familiar refrain of poor soil conditions and various weaknesses in the plants, and presuming that mature trees were resistant.

Denial of the burgeoning epidemic was encouraged by the variety of disease symptoms described from different locations. This led investigators to believe that they were dealing with a number of different fungi. Thomas Petch was the first to realize that the same species was killing rubber trees throughout the continent. This fungus was already known to mycologists:

it was called *Dothidella ulei*, after Ernst Ule, who collected it in the Amazon. It was later renamed *Microcyclus ulei*. *Microcyclus* was found wherever *Hevea* grew wild and, contrary to the growers' prayers, showed no age discrimination—it was an equal opportunity killer—attacking seedlings and mature trees with equal vigor. The disease was named South American Leaf Blight by Dutch plant pathologist Gerold Stahel in 1917.[26] Stahel was director of the Agricultural Experiment Station at Paramaribo in Suriname. For a plant pathologist working in the early years of the twentieth century there was no better job on earth. Leaf blight was the second fungal epidemic reported by Stahel: two years earlier he had penned the first scientific description of the cause of witches' broom of cacao. He was at the epicenter of the developing triumph of the fungi.

The blight begins, as do all of the diseases I've described so far, with the arrival of an infectious spore on the plant surface. *Microcyclus* produces two kinds of spores.[27] Its conidia—the type formed without any sexual get-together, by a single strain of the fungus—are quite large as spores go and are shaped like juggler's clubs (fig. 5.3). They stick to *Hevea* leaves, germinate, penetrate, feed, and distort the tissues. After a few days, the fungus reemerges from the bottom of the leaf and begins shedding the next generation of conidia. The spore-producing lesions have an olive color, and when enough of them bloom on a single leaf they coalesce into a continuous infectious sheet. By this time, the leaf isn't much use to the tree: it turns black and falls off. Premature leaf fall is a clear symptom of the blight. When the fungus develops more slowly, it has the opportunity to manifest itself in other ways besides its conidia. This stalling of its growth is obvious when the fungus is growing on older leaves, or during the infection of a rubber variety that has partial resistance to the disease. In these instances, spore-producing chambers called pycnidia have time to develop on the upper leaf surface. (*Microcyclus* pycnidia are similar to the structures formed by the chestnut blight fungus described in chapter 1.) The spores formed inside the pycnidia don't seem to be infectious, which begs the question of their function.[28] They probably serve as gametes, with spores from compatible pycnidia fertilizing one another like animal sperm cells and eggs. This kind of sexual interaction is common among fungi, and I'll explain how it works in rust fungi in the next chapter. For the sake of completing the life cycle of the leaf blight fungus, just assume that the birds and the bees do their stuff and mum and dad *Microcyclus* become proud par-

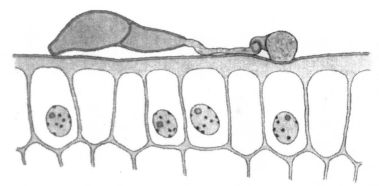

Fig. 5.3   Georg Stahel's 1917 illustration of the rubber blight fungus, *Microcyclus ulei*, infecting a leaf. The two-celled spore, or conidium, on the left has germinated on the leaf cuticle. A germ tube has grown a short distance over the leaf surface, and its tip has swollen to form a penetration structure called the appressorium. Infection proceeds when a peg grows from the base of the appressorium, pierces the cuticle, and enters the underlying cells of the leaf. From G. Stahel, *Bulletin Department van den Landouw, Suriname* 34, 1–111 (1917).

ents. Their offspring are ascospores, massive numbers of them, that are expelled from flask-shaped perithecia. (Again, just like chestnut blight.) By the time the perithecia develop, the leaf tissues between the upper and lower lesions are dead, forming dry discs that fall out leaving conspicuous holes. The sickened tree appears to have been scorched with a blowtorch and blasted with a shotgun. If you're a tire or condom manufacturer, take some deep breaths and go for a walk.

Both conidia and ascospores are infectious, but neither spore appears to remain viable for more than a few days after its release from the diseased leaf.[29] This precludes long-distance dissemination of the disease by wind and emphasizes the importance of humans in spreading the blight, either by transporting infected plants, or by acting as vectors when we carry spores on our clothing. Conidia have been isolated from the clothes and fingernails of visitors to plant nurseries infected with *Microcyclus*.[30] Dry air currents are ineffective at separating the conidia from leaves, but raindrops will puff them skyward in spectacular clouds. All of this fits with the natural, prehuman history of a disease that splashed around in the rainforests, infecting the widely spaced rubber trees, never threatening *Hevea* with extinction. This picture of a slow-moving fungus doesn't seem like a good strategy for a pathogen, and it isn't. But that's because I haven't given you the whole picture. *Microcyclus* was a suceessful tropical fungus long before we offered millions of susceptible plants in rubber plantations, because it infects several species of *Hevea* species besides Pará rubber. In other words, it had never suffered from any shortage of hosts, but plantation agriculture turned this microbe into a glutton.

Rubber production in Suriname was decimated by the blight. Entire groves of trees were destroyed and the fungus spread throughout the country. By 1916 the disease achieved epidemic status in British Guiana, and in Trinidad.[31] The height of the rubber tree canopy precluded disease control by spraying, and, in any case, the reservoir of the fungus on diverse wild *Hevea* plants in the surrounding forests suggested that fungicides would be useless. Pruning of infected branches was recommended, but the situation was hopeless. Production hurtled toward zero in the entire region. The falling price of rubber due to Asian exports was another strike against the growers in South America. At the beginning of the twentieth century, no more than 4 tonnes of rubber were produced in Southeast Asia. By 1910 this had risen to 100,000 tonnes.

It is worth underlining a trinity of mycological considerations, or laws, associated with growing any kind of crops anywhere. The first law states that *fungi find monocultures appealing.* A monoculture offers the microbe an ocean of prey, and every individual in the crop is identically susceptible, or resistant, to attack. The second law states that *crops grown outside their natural range are furnished with immediate protection from the fungal predators that kill them at home,* unless one is unlucky enough to convey the pathogens with their prey. A metaphor may help illuminate this concept: Imagine that a group of volunteers (or, more accurately, a flock of halfwits) charged with conveying rabbits to a petting zoo forgets to expel the foxes from the corners of the rabbit cages before loading them onto the truck. Finally, the third law serves as a damned if you do, damned if you don't warning for farmers: *crops grown outside their native range are threatened by all kinds of pests that occur in their new home.* Marshall Ward was one of the first scientists to acknowledge the problems with monocultures (chapter 3), and one might think that any intelligent person would have applied this knowledge to the natural rubber industry. But, as many joke, the French have always espoused a unique view of reality. In 1922, the botanist Paul le Cointe fed Brazilian nationalism with his pronouncement that "in the Amazon the Hevea is chez lui [the plant excels], thus equipped to resist enemies in the environment in which it has always won."[32] This statement has some validity when we consider the survival of a widely dispersed native plant, but not when it is transformed into a monoculture. *Hevea* may be in its element in Brazil, but only when one compares its situation to a bull in a bullring. It took a long time, and vast sums of American money, to make le Cointe's viewpoint die.

The history that I've recounted to this point stops just shy of the boom for rubber generated by Henry Ford. Production rose steadily in Southeast Asia, and the commodity price fell to the point that British-owned plantations became unprofitable. The price was boosted, and the plantations were saved, by the imposition of a government surtax that reduced exports from the colonies. American investment in Asia was limited, and instead became focused on Central America and the Carribean, where plantations were spared by the blight. Le Cointe wasn't alone in his optimism about Brazil. The low cost of land in the country, and its obvious suitability for *wild* rubber, seemed, albeit irrationally, attractive to Henry Ford. Following an encouraging survey of potential sites in the 1920s, the Ford Motor Company opened a massive concession there that became known as Fordlandia on the banks of the Tapajós River. Ironically, this was on the other side of the river from the site where Wickham had collected his seeds half a century earlier and broken the Brazilian rubber monopoly. Later, cognizant of the possibility of an outbreak of the leaf blight, trees that showed increased disease resistance were planted at a second site with richer soil called Belterra, close to Fordlandia. By 1935 *Microcyclus* was infecting nursery plants and cutting swathes through the juvenile plantations at Fordlandia. When the spores caught up with the supposedly disease-resistant clones planted at Belterra, they dropped their leaves too and perished under the tropical sun.

By 1940, 97 percent of the global production of 1.6 million tonnes of rubber came from Southeast Asia. The dependence of the allied nations upon Asian imports led to national crisis when the Japanese attacked Pearl Harbor. Within months, the majority of the rubber-producing countries were subdued by Japanese forces. The need for rubber had grown immensely in the pre-war decades: it insulated every electrical wire, and was an essential component in warships, airplanes, and tanks—each Sherman tank contained half a tonne of rubber. An American program of rubber conservation and recycling was introduced immediately, including the imposition of a 56 kilometer per hour (35 mile per hour) speed limit on highways. With the signal failure of the plantations in Brazil, tapping of wild rubber trees in the Amazon became crucial to the war effort, and the search for a method of synthesizing artificial rubber became a national imperative. The synthetic rubber industry was birthed from this wartime research, with the development of an array of polymers that mimicked some of the properties of the more versatile natural material.

Efforts to salvage the plantations at Fordlandia continued during the war. A technique called "top grafting" seemed promising. The method fused a disease-resistant crown (branches and leaves at the top of a tree) to the trunk of a rubber tree that yielded plenty of latex. The disease-resistant crown could be taken from other species of *Hevea*. The hope was that these Frankenstein-monster trees would yield rubber for a few years before they succumbed to the fungus. But by the end of the war it was clear that the plants chosen for crowns weren't as hardy as had been believed. The fungus killed them too. The Ford Motor Company had invested more than $20 million it its Amazonian plantations (approximately $250 million in today's currency), but by 1945 they provided a miserable yield of 115 tonnes of rubber. Beaten by a fungus, the company turned the plantations over to the Brazilian government for $250,000, which was sufficient to cover the severance pay for the Brazilian workers.

An Asian epidemic was forecast soon after the disease destroyed the plantations in Suriname and British Guiana. In 1922, for example, the colonial mycologist W. N. C. Belgrave said, "it does not require much imagination to picture the effect of this or any other leaf disease on the massed plantations of the Federated Malay States."[33] But decade after decade, *Microcyclus* didn't appear in Asia, and a century after the first epidemics it still hasn't emigrated far from South America. The contrast with the coffee rust's airborne escape from Ceylon is startling, but it's worth remembering that the major coffee-growing estates in Bahia remained disease-free for a century after the epidemic in Asia. The fragility of the spores of *Microcyclus* helps explain why the Asian rubber plantations established from Wickham's seeds have remained blight free, but I find it difficult to believe that this honeymoon can continue for much longer. If rubber blight does make its way to a single plantation in Africa, or worse to Asia, it could eliminate the bulk of natural rubber production within a few years. The climatic conditions in the majority of the rubber-producing areas of Asia are very similar to those in Brazil and the high-yielding plants grown in the region are highly susceptible to the fungus. There is no wiggle room for an optimist, because the repeated failure of plantations in South America in the twentieth century offers a lengthy and depressing test case for the world's rubber producers. Ethnobotanist Wade Davis offers a haunting evocation of the imagined export of the disease from Brazil:

> The day that haunts the rubber industry will dawn like any other. The sun, rising the length of Asia, slowly burns away the haze from the plantations along the South China Sea that are the source of 93% of the world's rubber. In fields the size of nations, shadows merge with the silver trunks of millions of identical trees, the most recently domesticated of all the major crops, one vast genetic clone spawned a century ago . . . The leaves, fresh and pliant a week ago, are withered and dry, blackened with lesions.[34]

Mentioning no names (look at the ones stamped on your tires), I am astonished by the *apparent* ostrich-like behavior of the rubber-manufacturing companies toward the disease. Conferences on *Microcyclus* have been organized in recent years (Michelin funded one in Brazil in 2004), but the scientific endeavor to understand the fungus appears frozen. I used the word apparent above, because I assume that the companies are funding studies that never see the light of day in the peer-reviewed scientific literature. Nevertheless, the lack of recent publications in the public domain is remarkable. The regular scientific literature on any topic is usually some reflection of the behind-the-scenes effort. Using the keyword *Microcyclus* in a database called Science Citation Index I uncovered only six papers on the fungus written in the last five years; "leaf blight AND rubber" unearthed a couple of additional publications, but this isn't impressive. Type "chestnut blight" into the same search tool and you'll crash your computer. There have been some good studies on *Microcyclus*, but there haven't been many of them.[35]

Today's rubber grower has a few practical options for combating the blight. Chemical controls have been developed in recent decades, and a variety of fungicides will control the fungus if the leaves are sprayed with a sufficient dose at the right time. These are effective in protecting young plants, but even a tractor-mounted pneumatic sprayer cannot soak the crown of a mature tree. Aerial spraying is the only viable method for dousing the tall trees of a productive plantation, but the cost of this soon overshadows the value of the latex. The development of *Hevea* clones with partial resistance to *Microcyclus* continue, as do long-term top-grafting trials.[36] A lower-tech approach hinges upon the discovery of "escape areas"—places where the climate supports *Hevea* but discourages *Microcyclus*. This method has allowed rubber production in parts of Brazil, but the growers in these places must often feel like shepherds surrounded by wolves.

Although Asian growers haven't spotted *Microcyclus*, I don't want to leave you with the impression that Wickham's introduction has enjoyed a disease-free century a hemisphere away from its homeland. Thomas Petch found plenty of root and stem illnesses to write about in his book on *The Physiology and Diseases of Hevea braziliensis* published in 1911, and many more diseases were recognized in succeeding decades.[37] The worst of these is caused by a bracket fungus called *Rigidoporus lignosus*. This is a basidiomycete, a member of the same taxonomic phylum as the coffee rust fungus and the mushroom that causes witches' broom of cacao. Unlike the little fleshy mushrooms of *Crinipellis*, however, this fungus sheds its spores from hard brackets that splay from the base of the rubber tree. Inside the tree, enzymes trickle from the edge of the feeding mycelium of *Rigidoporus*, dissolving the cells in the wood in a manner akin to the action of acids upon a corpse lying in a bath. *Rigidoporus* is a white-rotter, which, according to Brazilians, is an epithet made for Henry Wickham, and in textbooks means that its enzymes attack things like lignin and suberin in the wood, but leave most of the cellulose behind. The textbook picture is a bit simplistic. Although wood attacked by this fungus turns white, *Rigidoporus* destroys plenty of the cellulose in the walls of the cells that it penetrates in the early stages of the disease.[38] The fungus affects the rubber yield in the earliest phases of the disease by attacking the phloem tissues beneath the bark, which causes the latex to coagulate.

Like many other wood-rotting fungi, *Rigidoporus* moves from tree to tree by means of rootlike organs called rhizomorphs. Working in Malaya, plant pathologist Robert Napper (1907–1942) figured out that the fungus was a natural inhabitant of the forests that were cleared to make way for rubber.[39] Planters removed the trees that it had evolved to eat, but, fortunately for *Rigidoporus*, offered millions of *Hevea* plants in recompense. Napper developed methods for controlling the disease by removing older tapped-out trees before they could serve as a nest of infectious rhizomorphs. The biology of this fungus is similar to the dry rot fungi that destroy timber in buildings. In both cases, the fungus is a benign wood-decay microbe that attacks things that we care about when we rob it of its natural food.[40]

On a biographical note about Napper, he and his Dutch wife were killed in 1942 when their ship was bombed by Japanese aircraft as they were evacuated from Singapore. On a biographical note about the fungus he studied, the bracket fruiting bodies of *Rigidoporus lignosus* are no bigger than dinner plates, but those formed by its relative *Rigidoporus ulmarius* are more impressive: A specimen at Kew weighs almost 300 kg and is listed in *The Guinness Book of Records* as the biggest fruiting body in existence.[41] An American relative called *Bridgeoporus nobilissimus* gets almost as large as this in the rainforests of the Pacific Northwest. So vulnerable are the enormous brackets that their remaining locations are kept secret. Bracket fungi produce colossal numbers of spores. In the 1920s, the mycological mastermind A. H. R. Buller estimated that a single large bracket could shed 20 million spores every minute during a 6-month period of annual activity.[42] This corresponds to a total of 5 trillion spores per year. Many of the brackets are perennial, meaning that they develop a new layer of spore-producing tissue every year. The spores form on cells that line tiny tubes on the underside of the bracket. Once mature, they are then propelled from the interior surface of the tubes and fall under gravity where they are dispersed in the turbulent air flowing around the tree. The successive layers of tubes are clearly visible in a cut bracket, so it is possible to age the fruiting body by counting the layers, just as one ages a tree. But I digress.

Rigidoporus has become a particularly serious problem in the rubber plantations of West Africa, but in common with the precarious position of the industry in Asia, a single spore of *Microcyclus* represents an infinitely greater threat to African growers. Nothing else has the power to terminate the global flow of latex. The natural rubber industry supports the livelihoods of 30 million people. Natural rubber is an irreplaceable component of tires. Wholly synthetic tires work, but not very well, and then there are the multiple problems associated with our ever-increasing consumption of oil to manufacture those tires. And what about condoms? By robbing humanity of the uniquely protective barrier that natural latex offers to viruses, the danger posed by the rubber fungus may be far worse than any other microbe in this book.

CHAPTER 6

Cereal Killers

To catch the attention of students passing my laboratory, I posted the following question and answer on the door a few months ago: *Why study fungi?* Answer: The Romans invented a god of mildew, called Robigus, who was charged with averting crop diseases. Strictly speaking, this deity was supposed to concentrate on infections caused by rusts rather than mildews, but mycological knowledge was limited in ancient Italy. To rebuff crop damage, worshipers sacrificed a rust-colored dog (and a sheep, for good measure) at the annual festival of Robigalia on April 25. Based on this primeval wisdom, we might try executing someone with a fever to offset global warming.[1] Two thousand years have passed, but fungi remain the greatest threat to our crops. Science offers infinitely better prospects for changing this picture than any prayer, but only if you open this door.

Nobody seems to have read the notice. Undergraduates drifted by the lab, lost in their dreams of Wall Street, or of a private practice in cosmetic surgery. The science of mycology can be a tough sell. So when I get desperate for another lab assistant, I'll try some stronger medicine: a poster of hallucinogenic mushrooms. Within a day or so of hoisting this selective bait, scraggly undergraduates will flock to my lab like pigeons to breadcrumbs. The fungi will triumph.

୫ରୁ

The Romans created gods and goddesses on the slimmest excuse. Cloacina, goddess of the Cloaca Maxima, the system of sewers in Rome, is my favorite odd one. But in the case of Robigus, the ancients were responding to their dire need for a supernatural plant pathologist. There were lots of Romans to feed, and the fungi were wrecking too much of the annual wheat harvest. According to Ovid, the priest at Robigalia would ask Robigus to "spare the sprouting grain" and "forestall the destroyer." Unfortunately, Robigus was hearing impaired: The climate shifted toward a period of cooler and wetter weather during the first century A.D., and the frequency of rust epidemics increased. The resulting famine and social disorder are believed to have contributed to the eventual collapse of the Roman Empire.[2]

Robigus dissolved with his creators,[3] but his festival was incorporated into the Christian calendar. The Anglican Church continues to bless the crops on Rogation Sunday, or Rogationtide, which is celebrated in May. In many parishes in Britain this is associated with the annual Beating of the Bounds, when people walk the perimeter of their villages to mark their ancient boundary. In Drayton St Leonard, the bounds are soundly beaten before villagers restore themselves at a tea party or in the pub. As it happens, my mother circumnavigated the village this May morning as I am writing, relating the event to me over the phone. I suspect that the celebration in the Catherine Wheel is still in progress as I clatter away on my computer, far from the chinking of beer glasses.

Lucidity wasn't born with the demise of Robigus. Humanity would stumble through century after superstitious century before the nature of the microorganisms that destroyed the crops became clear. The blights, rusts, and rots of the previous chapters served as examples of relatively recent conflicts with fungi, but the rusts and smuts of cereals have afflicted us since the first Neolithic farmers planted cereals 10,000 years ago and noticed charcoaled ears and black-streaked leaves. Rust and smut infections of cereals remain a major agricultural concern for the obvious reason that these domesticated grasses account for more than half of our food supply today. Rice is resistant to these fungi, but it succumbs to other plagues that I'll discuss in the final chapter; that leaves 1.5 gigatonnes of grain threatened by rust and smut fungi.[4] They cause billions of dollars of losses every year, and billions more are invested in protecting crops from infection with fungicides. Our understanding of these diseases reaches back to the birth of the study of the fungi, and this chapter is concerned with the story of this unfolding intelligence.

The study of fungal diseases begins with the bunt, or stinking smut of wheat, that transforms the grain into chocolate-brown or black powder with the odor of rotting fish (fig. 6.1). The English agriculturalist Jethro Tull wrote in 1731 that "Smuttiness is when the Grains of Wheat, instead of

Fig. 6.1   Bunt of wheat showing dark bunted grains cradled by the dry glumes of the diseased wheat ear. From T. Milburn, *Fungoid Diseases of Farm and Garden Crops* (London: Longmans, Green & Company, 1915).

Flour, are full of a black stinking powder."[5] Tull didn't know it, of course, but the stink was due to the production of a compound by the fungus called trimethylamine (C_3H_9N). In Tull's time, the cause of the smuttiness had been ascribed to dangerous mists (which isn't a bad guess because individual smut spores are invisible without a microscope), the full moon, insects, faulty grain, and to the action of raindrops as lenses that burned the leaves with sunlight. But during the next century, the scientific study of plant diseases began to unfold from the work of a pair of scientific pioneers from southern France: Mathieu Tillet, working in the 1750s, and Bénédict Prévost, in the early 1800s.

Not much is known about Tillet besides his revolutionary work on stinking bunt of wheat.[6] His researches were stimulated by a competition organized by the Academy of Arts and Sciences of Bordeaux. The academy offered a prize for the best dissertation on the cause of the bunt and how it might be treated. Tillet's approach to the problem was unorthodox. He chose to test clearly stated ideas by performing experiments, furnishing the results of the experiments for his readers, and offering his own interpretation. Tillet was following the tenets of the experimental method promoted by Francis Bacon. In Tillet's time, French science was steeped in a philosophical tradition that encouraged endless speculation and discourse over practical investigation. In Voltaire's *Letters on England* (1733) the Académie de Sciences was cast as a pompous and ineffectual organization that did little but print endless addresses that affirmed the greatness of aristocrats and clergymen.[7] Voltaire celebrated the genius of Bacon and Isaac Newton and lauded Edward Jenner's vaccination against smallpox, while decrying the medieval superstitions that quashed scientific progress in France. (He was impressed by his countryman Descartes, though noting that the great philosopher had fled to Holland in an ultimately futile quest to be taken seriously.) Not surprisingly, Voltaire's book sold well in Britain but was banned in France.

Against this background, Tillet's reliance on experiments was daringly novel, and so he took pains in the introduction to his dissertation to placate his readers: "I soon realized that after devoting myself to helpful speculation it was still necessary to have recourse to experiments." Tillet planted an experimental wheat crop that he divided into 120 separate subplots in which he tested the protective value of different combinations of manure, salt, lime, and saltpeter (potassium nitrate) on clean seeds and on seeds exposed to the black dust from bunted plants. Tillet included the appropriate controls, including plots without any manure, and sowed groups of subplots on different days to examine the influence of weather conditions. This complex experimental design allowed him to evaluate the effects of multiple combinations of variables on the development of the bunt. Happily, the results were very straightforward: seeds exposed to the black dust grew into diseased plants; the majority of the clean seeds grew into healthy plants, and seeds treated with lime produced healthy plants even when they had been exposed to the powder. This monumental field trial conducted in 1751, and strengthened by follow-up experiments in 1752 and 1753, produced a giant leap in understanding of the bunt: "the common cause, the abounding source of bunted wheat plants resides in the dust of the bunt balls of diseased wheat." In addition to proving the innocence of the full moon, Tillet had also formulated the first method for protecting cereals from disease.[8]

Prévost was to take the next major step in bunt research by building on Tillet's work and studying the causal fungus with the aid of a microscope.[9] Prévost was a professor of philosophy and self-taught scientist who worked in the city of Montauban, north of Toulouse. Like Tillet, his investigations on smut were stimulated by an invitation extended by a local scientific organization to its members—in Prévost's case this was the Society of Montauban. This practice is like the annual solicitation of grant proposals on specific topics advertised by the United States Department of Agriculture, though far more money is involved in today's appeals for critical research. Placing one of the bunted grains into water, Prévost watched the dust pour from a rip in its wall "like a descending smoke." The unending flow of particles into the water was a mesmerizing demonstration of the infectious potential of the fungus. Where Tillet had seen dust, Prévost was able to gaze upon the individual spores through his microscope. He calculated that each grain contained one million of them. Mixing a few bunts in a tumbler of water, he then studied the germination of the spores by withdrawing drops at various times. The spores put forth stalks, and each stalk bore a tuft of leaflike plumes—an aigrette. (Each of the plumes is a type of spore called a primary sporidium.) A little later, small banana-shaped bodies developed from the side of the plumes on short, wiry stalks. Prévost believed, correctly, that these bodies were fruits, or spores, produced by the plumes. (The banana-shaped bodies are the secondary sporidia, or ballistospores, of the smut and are dispersed by wind.) Making the best of his pitiful microscope, Prévost provided engravings of the developing spores in his 1807 report. Copies of the original report are very rare, but you can get an idea of the quality of the drawings in figure 6.2. When I looked at an original copy at the Lloyd Library in Cincinnati, I was struck by the simplicity of the drawings: Prévost was no artist. But to his contemporaries, the book was sacred. Sir Joseph Banks, Cook's naturalist, prized his copy, and Anton de Bary was frustrated by his inability to obtain one in the 1850s.[10] The scratchy illustrations are among the most important images in the history of science.

Prévost went much further than illustrating the spores; he was the first person to offer experimental proof that a particular microorganism caused a specific disease. Tillet fans can't claim that their hero had achieved the same thing much earlier, because without seeing the fungus with a microscope he could only link the disease to the action of the mysterious dust. Prévost demonstrated that the spores of the bunt fungus germinated in soil in the same way that they behaved in water. He then surmised that the microbe infected wheat seedlings and grew up the developing stem into the flowers. Once inside the flowers, the fungus deformed the reproductive organs and converted the developing grain into spore-filled bunt balls.

Finally, in support of his candidacy for scientific sainthood, Prévost discovered a new remedy for the disease. Since Tillet's work, other investigators had shown that arsenic and mercury were very effective at preventing the bunt, but widespread use of these poisonous chemicals had some

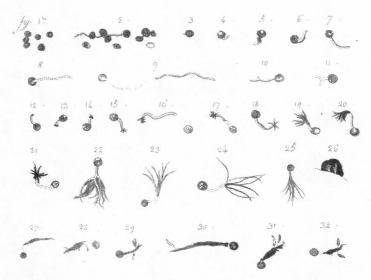

Fig. 6.2 Bénédict Prévost's classic illustrations of the spores of the bunt fungus, *Tilletia caries*. The drawings in the two rows at the top of the plate (1–11) show the germinating bunt spores. The next pair of rows (12–26) show the developing plume of elongated spores, or primary sporidia. The bottom row (27–32) shows secondary sporidia, or ballistospores, developing from the sides of the primary sporidia. From B. Prévost, *Memoire sur la cause immediate de la carie ou charbon des bles, et de plusieurs autres maladies des plantes, et sur les preservatifs de la carie.* (Paris: Chez Bernard, 1807).

obvious drawbacks. Serendipity blessed Prévost. During his investigations in Montauban, he had been shown a copper colander that was used to steep wheat seeds in a vat of lime and sheep urine. Crops grown from seed treated in this manner were remarkably smut-free, the effect being attributed to the lime treatment recommended by Tillet. Prévost soon showed that it was the copper leaching from the colander, rather than the lime, that offered the greatest protection for the seeds. Using his microscope, he found that highly diluted solutions of copper sulfate were very effective at suppressing the germination of the spores, and Tillet-style field trials proved the effectiveness of the new fungicide.

It's difficult from our contemporary perspective to grasp the magnitude of Tillet's and Prévost's achievements because we take the power of experimentation for granted. Imagining a time when information was sought largely by speculation guided by religious fantasy, is tough in the context of modern scientific methods. Our appreciation of these early struggles to understand plant diseases is also compromised by our familiarity with microscopic organisms. Consider the background to Prévost's discovery. Fungal spores were first mentioned by Giambattista della Porta, who offered vague descriptions of tiny black seeds in his *Phytognomonica*, printed in 1588. It isn't clear which fungi he studied in the century before the invention of the microscope, but it seems likely that he looked at the dust particles emanating from puffballs and slime molds and surmised their function.[11] Hooke, Malpighi, and Leeuwenhoek published illustrations of fungi using their seventeenth-century microscopes, but another century passed before the function of fungal spores was understood. Joseph Pitton de Tournefort drew germinating mushroom spores in 1707, and in 1729 the Florentine scientist Pier Antonio Micheli demonstrated that spores produced new generations of the organisms that shed them. Micheli's findings, published in his masterpiece, *Nova Plantarum Genera*, should have flushed the classical theory of spontaneous generation into the Cloaca Maxima. But this ancient folly died hard. A few traditionalists even continued to question whether fungi were living things at all. The prize for utter foolishness during this part of the history of mycology goes to an imaginative German named J. S. T. Frenzel, who proposed, in 1804, that fungi were birthed by shooting stars.[12]

The importance of Prévost's work is also interesting in light of its genesis during the French Revolution. Crop losses were very high in the 1780s, leading to the scarcity of flour, which was one of numerous sources of disquiet among the poor. In her book *Poisons of the Past*, historian Mary Kilbourne Matossian develops a case for the impact of the ergot fungus on the morale of the peasants.[13] Ergot fungus, *Claviceps purpurea*, grows on rye plants in wet years and generates a dreadful melange of vasoconstricting toxins, plus the original version of LSD. (I discussed ergot in my book *Mr. Bloomfield's Orchard* and will not delve further here.) The combined impact of food shortages and an epidemic of paranoia (both of fungal origin) would certainly have contributed to the unrest, and the insensitivity of Marie Antoinette, and other aristocrats, to the plight of revolting peasants was one of the justifications for separating heads from necks. Since the bunt, or stinking smut, was a major contributor to persistent crop losses in France, the plea for research by the Society of Montauban in 1797 was very timely. But despite its practical value, Prévost's work was overlooked until its rediscovery by Louis-René Tulasne and his younger brother Charles half a century later. The Tulasnes embarked upon a systematic study of the rust and smut fungi in the 1840s. They distinguished between the smut fungus that caused the bunt (or covered smut), and named this *Tilletia caries* (after Tillet of course), and other smuts that obliterated the wheat ears before they were harvested.[14] The second type of infection was known as loose smut and was common among a variety of grass species. The Tulasnes recognized that different species of smut fungi caused loose smut in different grasses, and they assembled these fungi under the generic name *Ustilago*. This classification has persisted to the present day.

Shifting our attention across the Atlantic, stinking smut wasn't a tremendous problem in North America until the prairies were converted into an ocean of wheat in the late nineteenth century. In 1890, between 25 and 50 percent of the Kansas crop was lost to stinking bunt, and the fungus ruined harvests all the way to eastern Washington and across the Canadian wheat belt.[15] Clouds of smut teliospores erupting from the bunted grains became highly charged in the bone-dry air, and these dust clouds exploded with alarming regularity as teams of 20–40 horses dragged harvesting machines through the diseased fields. The explosions were fired by the combustion of the odorous trimethylamine, which is, unfortunately, as flammable as it is smelly.[16] Prévost's copper sulfate treatment was used to control the disease, but a method that worked well on small European farms wasn't an efficient treatment for the mountains of seed destined for the prairies. Only seeds steeped in copper sulfate solution and then thoroughly dried before planting would rebuff the fungus. This laborious process was eliminated around the time of the First World War by the introduction of a powdered preparation of copper carbonate. This became very popular, and organic fungicides containing mercury proved even more successful. Widespread use of these compounds foreshadowed today's reliance upon an armory of chemical fungicides to control the major cereal diseases.

Smuts remain an agricultural problem, pairing away at the slim profit margin of modern cereal growers. Stinking smut accounts for the loss of a few percent of the American wheat crop every year and sometimes flourishes to epidemic levels. Operators of grain elevators refuse to accept grain with the unmistakable fishy odor, or at least pay far less for a delivery spoiled by "smutty" grains. Besides the depreciation of contaminated wheat, the propensity of dry spores to ignite harvesting and threshing machinery poses a similar threat in today's grain elevators—or would do so if enough of them passed the inspectors. Smut infection is controlled by planting seeds dusted with color-coded fungicide mixtures: treated seed corn is easily identified by its bright red hue. Despite the availability of modern fungicides, a few smut species continue to be exceedingly important players in global agriculture. *Tilletia controversa* causes dwarf bunt of winter wheat that reduces plant height and substitutes smelly spores for grain. The disease is most problematic in locations where the soil temperature fluctuates around freezing for many weeks. This has been of particular concern to Chinese growers and prompted an import ban on American wheat that did not carry a smut-free guarantee (the ban was lifted in 2000).

Corn is second only to wheat in the ranking of crops grown in the United States. Annual production of this commodity is staggering: 9 billion bushels are harvested every year from 279,000 square kilometers of farmland;[17] for comparison, the land area of the United Kingdom is 245,000 square kilometers. Close to half the crop is genetically modified, or augmented, with the bacterial Bt toxin that kills insect pests, and much is also furnished with genetic resistance to herbicides. These modifications do nothing, however, to protect the plant against corn smut, caused by *Ustilago maydis*, that replaces kernels with spore-filled galls or tumors (fig. 6.3). The fungus can grow as a yeast, with cigar-shaped cells that proliferate by forming buds, or in the form of hyphal filaments. This behavior is termed *dimorphism*. The yeast form allows the smut fungus to spread as a saprobe on the surface of dead corn tissues. The switch to invasive filaments occurs when two compatible strains of the smut mate on the surface of living corn plants. The smut then enters the plant through its stomatal openings, or via its silks. Corn silks are the dangly female flower parts that capture pollen. By growing through the silks, the fungus enters the plant's ovaries, which is how the infected flowers develop into bags of smut spores rather than corn kernels. The fungus affects the balance of hormones within its host, which contributes to the tumorous swelling of the plant tissues, and *Ustilago* converts itself into billions of brown-black teliospores within each of these swellings.

An illustration of corn smut in a sixteenth-century manuscript on the Aztec culture is the oldest portrait of a plant disease caused by a fungus of unambiguous identity.[18] The Aztecs, and, in all probability, the Incas and Maya as well, were very familiar with the smut and probably enjoyed eating it. Today, the galls are the source of the Mexican delicacy called *huitlacoche*. If you type "corn smut recipe" into your web browser, you'll capture plenty

Fig. 6.3 Symptoms of corn smut caused by *Ustilago maydis*. The kernels in the upper part of the cob are swollen and filled with smut spores. From O. Brefeld, *Untersuchungen aus dem Gesammtgebiete der Mykologie. Fortsetzung der Schimmel- und Hefenpilze. XI. Heft: Die Brandpilze II. (Fortsetzung des V. Heftes.) Die Brandkrankheiten des Getreides* (Münster i. W., Germany: Verlag von Heinrich Schöningh, 1895).

of lovely sounding dishes including cream of huitlacoche soup, crepes filled with corn fungus, and so on. Despite such endorsements, I have never felt eager to dine upon these fungal tumors, and have, so far, avoided the experience.

Given the interest in huitlacoche (some Mexican farmers specialize in deliberately infecting their crops) the incorporation of smut spores into the corn that is milled into breakfast cereals might seem like a bonus. But while our cornflakes may be smutty, they are never very smutty. Federal guidelines in the United States stipulate that to qualify as Grade 1 corn, no more than 3 percent of the kernels can be blemished by smut, insects, and other sources of material damage.[19] Incidentally, for international readers, it may be useful to clarify that cornflakes are made from corn (maize or *Zea mays*) rather than wheat. Outside the United States, "corn" also refers to other cereal grasses. In the United Kingdom, for example, corn is wheat rather than maize. (Indeed, I was surprised to discover, as an adult, that cornflakes had nothing to do with wheat.)

The smut fungi are identified as the *Ustilaginomycetes* in current scientific parlance and comprise more than 1,400 species.[20] Besides the cereal

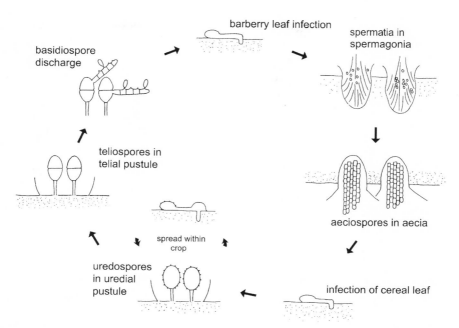

Fig. 6.4   Life cycle of the rust *Puccinia graminis*. From N. P. Money, *Mr. Bloomfield's Orchard: The Mysterious World of Mushrooms, Molds, and Mycologists* (New York: Oxford University Press, 2002), with permission.

crops, victims of the smuts are, with a couple of exceptions, limited to the flowering plants, mostly grasses and sedges. Their relatives, the rusts, or *Urediniomycetes*, number at least 7,000 species. These fungi attack flowering plants, conifers and other cone-bearing plants, and ferns. Because many plants are troubled by their own personal rust fungus, and because very few scientists specialize in documenting these relationships, the actual number of rusts is thought to be much higher, probably between 10,000 and 60,000 species.[21] The combined chunk of microscopic biodiversity is related to the mushroom-forming fungi, grouped in the *Hymenomycetes*, and all three classes—the smuts, rusts, and mushrooms—constitute the taxonomic phylum called Basidiomycota.

Compared with the dimorphic life cycle of *Ustilago*, the behavior of the rusts that attack cereal crops is immeasurably more complicated. Let's look at the workings of *Puccinia graminis*, the fungus that causes black stem rust of wheat. The complexity arises because, in addition to attacking the crop plant, *Puccinia* also infects the leaves of a thorny bush called barberry (fig. 6.4). To enable spread among its pair of hosts and to ensure its survival during winter, *Puccinia* produces four types of spore: uredospores, teliospores, and basidiospores on wheat,[22] and aeciospores on barberry. Finally, sexual reproduction involves sex cells called spermatia (some refer to these as a fifth spore type) that are moved around on infected barberry leaves by insects. As I said in *Mr. Bloomfield's Orchard*, this is the Fabergé egg of life cycles. It took centuries to unravel this story, but I'll summarize the history in a couple of paragraphs.

The Italians deserve a round of applause for their work on rusts in the eighteenth century. Giovanni Targioni-Tozzetti had been a student of Micheli in Florence. After his teacher's death in 1767, Targioni-Tozzetti pursued investigations on rusts and smuts with the stated aim of providing relief for the poor in Tuscany and other areas suffering from serious crop losses. His etchings of rust spores were awful (his mentor would have flogged him), but he came to a surprisingly clever conclusion: "the rust is an entire, very tiny, parasitic plant that grows nowhere except inside the skin of the wheat."[23] This was a remarkably prescient statement about the biology of this fungus, capturing the obligatory, or biotrophic, nature of the rust's parasitism long before anyone began to make sense of its life cycle. He hit upon the truth by painstaking microscopic study of the intimate physical connection between the rust and the tissues of the wheat plant. His contemporary, Felice Fontana, offered reasonable illustrations of the stalked teliospores responsible for the black wounds on the wheat leaves and stem, and perceived that these were interspersed by reddish "eggs." These eggs were the uredospores of the same rust, but Fontana came to the simpler conclusion that "two sorts of mildew" were affecting the same plant.[24] Time passed, and the two sorts of mildew were accorded different names until investigators recognized that these were two forms of the same thing. Large offers a rich telling of this part of the story in his *Advance of the Fungi*, and I won't repeat it here. But I do want to quote from the writings of the English botanist Joseph Banks in 1805, because his contributions to defrocking the wheat rust prefigure everything that followed:

> It has been long admitted by farmers, though scarcely credited by botanists, that wheat in the neighborhood of a barberry bush seldom escape the Blight. The village of Rollesby in Norfolk, where barberries abound, and wheat seldom succeeds, is called by the opprobrious appel-

lation of Mildew Rollesby . . . Is it not more than possible that the parasitic fungus of the barberry and that of the wheat are one and the same species, and that the seed is transferred from the barberry to the corn?[25]

Banks' insights are improbably brilliant. He had solved much of the life cycle half a century before Anton de Bary, whom most authors credit with the "discovery" of the life cycle. Banks is rarely mentioned by plant pathologists.[26] Heinrich Anton de Bary (1831–1888) was an exceedingly influential scientist who taught at the University of Strasbourg (fig. 6.5).[27] His first book, *Die Brandpilze*, on rusts and smuts was published when he was at medical school and only 22 years old.[28] Research by De Bary in the mid-nineteenth century, along with the work of the Tulasne brothers, was instrumental to the slow unfolding of the life cycle of the rusts. The great contribution of the Tulasnes was to trumpet the fact that a single rust species produced a succession of different spore types.[29] The rusts offered a dramatic example of a wider fungal phenomenon that they called pleomorphy, later celebrated in the brothers' three-volume magnum opus, *Selecta Fungorum Carpologia*, published in the 1860s.[30] De Bary "put the pieces of the jigsaw together" by inoculating wheat and barberry plants with the uredospores, teliospores, basidiospores, spermatia, and aeciospores.[31] Using

Fig. 6.5   Anton de Bary. From G. C. Ainsworth, *An Introduction to the History of Plant Pathology* (Cambridge, UK: Cambridge University Press, 1981), with permission.

this method, De Bary learned that the uredospores only infected wheat and settled the function of the other spore types, but the operation of the spermatia remained a mystery until the twentieth century.[32] But is it fair to say that De Bary solved the life cycle as many historical accounts have claimed? I'm not convinced by this attribution of breakthrough; the *Reader's Digest* version of this story is hollow. Science proceeds by discovery building upon discovery ad infinitum, and this was certainly true of the rust life cycle.

There is a difference between a scientist like De Bary, who is gifted with intelligence, thoroughness, and vision, and someone of pure genius. For historians of science this distinction is of fundamental importance. De Bary's work on the wheat rust and potato blight (discussed in the next chapter), extended Pasteur's assault upon spontaneous generation (based on work with bacteria) to the fungi.[33] De Bary brought clarity to the study of fungi and fungal diseases of plants, marshaling the power of careful experimentation to hammer out a synthesis of valid preexisting ideas while dismantling a lot of shoddy science. His textbook, *Morphologie und Physiologie de Pilze*, published in 1866, marked the beginning of the modern study of the fungi.[34] Besides his work on fungal life cycles, De Bary is also recognized for his assertion that blue-green algae were bacteria (they are known as cyanobacteria today), for demonstrating that yeasts are fungi, and for coining the term "symbiosis" in 1879, for "the living together of differently named organisms."[35] The scope of his work was astonishing, but he didn't develop truly novel concepts or revolutionary views in the vein of Mathieu Tillet and Bénédict Prévost. Indeed, if Nobel prizes were posthumous, De Bary might not make the cut, but Tillet and Prévost should. Failing this, the French can claim, in my opinion, to have invented plant pathology.

De Bary described the life cycles of species like *Puccinia graminis* as *macrocyclic* because there are so many spore-generating steps, and *heteroecious* because they involve a pair of alternate hosts.[36] How might these complex relationships with plants have developed? The first of many possible answers suggests that *Puccinia's* forebears worked by infecting barberry long before the evolution of grasses, and then filled in the rest of the life cycle to make use of this new food source. This would have required the ancestor to produce spores capable of reinfecting barberry, which is something that today's pathogen cannot do: aeciospores of *Puccinia* are shed to infect cereals and are useless against barberry. A second idea is that today's multiple spore-producing stages reflect their origin from a pair of ancestral parasites, one that infected barberry, the other that attacked cereals. Initially, both fungi may have evolved, during the course of millions of years, from benign leaf inhabitants to invaders of the living tissues of their respective hosts. At some point, according to this dual-species theory, they met on one another's plants and found that cohabitation, and, later, mingling of cell contents and genes enhanced their reproductive success. The hypothetical advantage of fusion would be that the availability of two plants invariably offered a more consistent food supply than any single host.

The dual-species idea has been encouraged by some recent research concerning a rust called *Tuberculina persinica*. Rather than infecting plants, this fungus lives among the colonies of other rusts while their rust hosts infect dandelions and anemones. Amazingly, *Tuberculina* fuses with the cells of these rusts and injects its nuclei through channels called fusion pores.[37] Eventually, *Tuberculina* produces spores within the spore-producing cups, called aecia, of its host. The host aeciospores fly off in search of fresh plants and are pursued by the *Tuberculina* spores. The behavior of *Tuberculina* is interesting in relation to the dual-species idea because it shows that closely-related rusts can mix their cell contents, and even swap nuclei, without killing one another. Such cooperation, or at least tolerance, is an essential prerequisite for the fusion of two life cycles.

�֍

Contemporary research on rusts concentrates on methods of control. Chemists design, test, and troubleshoot new fungicides, and plant breeders develop new wheat varieties (or cultivars) in a never-ending search for resistance to the constantly evolving pathogens. Genetic engineers combine these approaches by creating new-fangled cereals that produce their own fungicides, harvest their own grain, and package themselves into cereal boxes. This is, for now, a mild exaggeration, but you get the point about the straightforward "take-no-prisoners" strategy of genetic engineering. As I have explained earlier in the book, pathogens are engaged in an arms race to overcome the continually evolving defenses of their hosts. We accelerate this process by breeding and engineering new crop varieties. At the moment, wheat cultivars are remarkably effective at excluding *Puccinia graminis*. They are blessed with genetically encoded sensitivity to the most widespread varieties (or races) of the rust, and they seal off the infected entry points as soon as they appear. But this comfortable situation is unlikely to persist for long.

For heteroecious rusts, one might be forgiven for thinking that the odds of beating the disease are doubled. Shouldn't black stem rust of wheat be beaten by eradicating its alternate host, the barberry bush? It can be argued that the disease would never have plagued American wheat crops if immigrant European farmers hadn't imported barberry in the first place.[38] The introduction of the barberry wasn't an instance of deliberate bioterrorism. The plant was imported because it was useful as a hedge plant to protect livestock, and its red berries were used in jellies and to make wine. But even before the Revolutionary War it had become clear that the bush was bad for the grain, and programs of barberry eradication were implemented in Massachusetts, Connecticut, and Rhode Island.

Once the details of the collaboration between *Puccinia* and barberry were resolved, the practice of barberry removal gained the luster of scientific logic. As the North American wheat belt expanded, the losses due to fungi climbed into hundreds of millions of bushels per year, and a national policy of barberry annihilation was initiated by the U.S. Department of Agriculture during the First World War. "The common barberry is an outlaw in Colorado, Illinois, Indiana, Iowa, [etc.]" wrote plant pathologists in one of many *Farmers' Bulletins* on the disease, "Kill the barberry now and help to protect the grain crops of the future."[39] The Farmers' Bulletins and other leaflets distributed by the USDA were superbly detailed publications, offering clear descriptions of the complicated life cycle of the rust, and reporting the incredible numbers of infectious spores shed from a single barberry bush (64 billion). The recommended method of killing barberry was to pour salt or kerosene around the base of each bush to ensure the death of its root system. The method had proven somewhat successful in some European countries, but the scale of North American agriculture presented a substantially bigger challenge. Nevertheless, the federal program reduced crop losses in many areas and wasn't formally terminated until the 1970s.

Some experts believe that the program was cancelled too soon and are calling for renewed vigilance against the bush, which is reestablishing itself from its long-lived seeds. The shelf-life of a wheat cultivar is often limited to a few years, but the fact that *Puccinia* has lost a great deal of its barberry love-nests means that it cannot engage in sexual reproduction and, therefore, is stuck in an evolutionary hole. Without sex and recombination of its genes, it cannot evolve a way around the plant's defenses, which is good news for cereal growers. If you're following the logic here, the importance of a barberry rebound should be obvious: more barberry means more sexual reproduction, and more sexual reproduction means more virulent pathogens. Even if barberry were completely eradicated from North America, however, black stem rust would remain a problem. First, the uredospores are capable of overwintering, even in places with extremely cold winters like North Dakota, which circumvents the need for the sojourn on barberry before infecting the spring cereal crop. This problem will worsen, of course, if global warming transforms Fargo into the next Palm Springs. Second, and still worse, infectious uredospores can be blown hundreds of kilometers north from Mexico.[40] Having developed from rust pustules that grew on popular cereal varieties in Mexico, the migratory spores may someday possess the key to destroying the monocultures of different varieties undulating across the prairies. The continuing success of this fungus seems as certain as the job security of the agricultural researchers developing new cereal varieties.

Modern agriculture has succeeded in limiting the potentially devastating effects of the rusts and the smuts, but none of the cereal plagues has been eradicated. Keeping these fungi at bay is costly, both in the immediate term because fungicides are expensive, and, in the longer term because some of these compounds are flushed into our groundwater and degrade very slowly.[41] The rusts and the smuts (two ancient groups of basidiomycete fungi that do nothing else but infect plants) have triumphed in the sense that they have never gone away.

I often run along a path through the woods beside a wheat field. In summer, this Ohio farm makes me think of Van Gogh's paintings. If I didn't know he painted it in Provence, I'd swear his last work, *Wheat Fields with Crows* (1890), was spread madly onto canvas from this spot. Even those diabolical crows are here. When thunder rumbles across the border from Indiana, the dry leaves rustle in the building wind. And with the undulations of these amber waves, little clouds of spores puff from blackened ears. The chaos continues.

CHAPTER 7

Potato Soup

"Few subjects have attracted more attention, or have been more variously canvassed, than the malady with which Potatoes have been almost universally visited during the autumn of 1845," wrote the Reverend Miles Joseph Berkeley in 1846 (fig. 7.1).[1] Since my student days, I have associated Berkeley with the history of mycology and plant pathology, but a biographical essay by Stefan Buczacki convinced me that I should take a closer look at the great man's work.[2] To this end, I photocopied the reverend's article on the potato blight from the 1846 volume of the *Journal of the Horticultural Society of London* in the Lloyd Library in Cincinnati, and set off for my customary pint at Nicholson's pub. Half was read while I took the shortcuts through the alleyways of downtown, my nose in the fluttering pages, the rest while perched at the bar. It is a marvelous read.

Faced with an agricultural malady of unparalleled sweep ("the whole of Western Europe, from Norway to Bordeaux, seems almost equally to have suffered," p. 10), Berkeley set about a systematic analysis of what he saw as the most probable cause of the potato murrain: a parasitic fungus. With admirable sensitivity to other authors, he dismissed their theories about electrical influences, unusual weather, invisible insects, the use of animal manure, and the innate degeneracy of the potato, and concentrated on the effect of a single, microscopic disease-causing agent. Berkeley's prescience becomes evident midway through the essay: "The decay is the consequence of the presence of the mould, and not the mould of the decay . . . The plant then becomes unhealthy in consequence of the presence of the mould, which feeds upon its juices" (pp. 23–24). Berkeley was forcing mankind up a steep flight of stairs, leaving lots of old, redundant thinking behind. He went on to explain that although weather conditions affected the development of the

119

Fig. 7.1   Miles Berkeley. From *The Graphic* (November 15, 1873).

disease, rainfall didn't cause the blight. The fungus was "the immediate cause of the destruction." Berkeley's article expressed what others had concluded, but he framed his argument in such a cogent fashion that no logical person could continue to dispute the facts about the nature of the blight.[3] The certainty of Berkeley's dissertation underscores the stain upon the history of plant pathology left by Jakob Eriksson's ridiculous mycoplasm theory (chapter 3).

The potato fungus was named *Botrytis infestans* in the 1840s, and became *Phytophthora infestans* when De Bary renamed it in 1876.[4] A related species, *Phytophthora palmivora*, is one of the causes of black pod of cacao. As I explained in relation to the cacao disease in chapter 4, Phytophthoras are oomycete water molds; not fungi, exactly, but microbes that behave like them. I'll come back to this distinction shortly. Potato blight becomes apparent when spots develop along the leaf margins; these darken and spread, and the pathogen emerges on the underside of the leaves as glistening halos of white mycelia that cast spore-filled sporangia into the air (fig. 7.2). The pathogen likes everything about its host, attacking potato leaves, stems, and tubers (fig. 7.3). In wet weather, all of the leaves can succumb to the infection, and the stems wither. The farmer is left with fields of foul-smelling compost. (I had hoped that Oxford University Press could provide readers with a scratch-and-sniff patch at the bottom of this page, but my

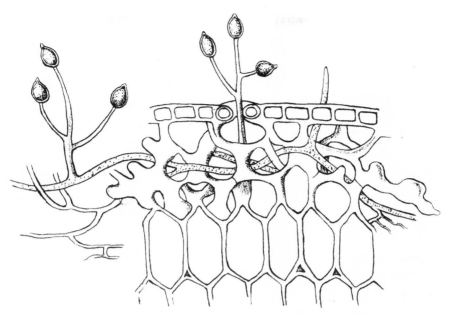

Fig. 7.2   Berkeley's illustration of the potato blight pathogen *Phytophthora infestans*. From *Journal of the Horticultural Society of London* 1, 9–34 (1846).

Fig. 7.3   Berkeley's illustration of infected potato tubers. From M. J. Berkeley, *Journal of the Horticultural Society of London* 1, 9–34 (1846).

editor opted for the verbal description—sorry.) The infected tubers become similarly offensive, displaying brown streaks that penetrate the vegetable's flesh and then turn it into soup.[5] The souping process, or "wet rot," is aided by other microorganisms, particularly bacteria, that exploit the dying tuber. Unlike the culinary interest in corn smut, I am not aware of any culture that covets this disease and eats the liquified tubers.

The disease is spread by the sporangia that form beneath the leaves. These can be splashed short distances by raindrops or carried hundreds of kilometers by wind. During warm weather the sporangia germinate like any other kind of spore, protruding a germ tube that extends and branches to form a new colony. But in cooler weather, when temperatures fall below 15° C (or 59° F), the development of the sporangium is more complicated. First, a plug at one end of the lemon-shaped sporangium inflates to form a thin-walled balloon called the vesicle. Meanwhile, the contents of the sporangium have been cleaved into a clutch of swimming zoospores. These spill into this vesicle. Finally, the vesicle bursts, and the spores swim away (fig. 7.4). This process of sporangial emptying, or discharge, is accomplished in a few seconds. If there were a *New York Times Review of Pathogens*, the critic would assuredly describe this as a display of cunning bravura and an emblem of the pathogen's venom toward the artless potato.

Spore discharge is important for the pathogen because the swimming spores can move through water on the surface of the potato plant and can also make their way through waterlogged soil and infect the tubers directly. On the leaf surface the zoospores jiggle around for a while (they can't swim beyond the slick of water that encompasses them), then they settle down, shed or retract their flagella, turn into a cyst, and stick to the leaf. Next comes cyst germination. A germ tube extends from the cyst, travels a short distance over the bumpy terrain of the leaf, then inflates at its tip to form an infection platform or appressorium. Finally, an infection hypha pushes downward from the underside of this appressorium, which is the region fixed to the leaf, and pierces the underlying cuticle and epidermal cell (fig. 7.5A). The colonization of the potato plant has begun.

There is an endless cycling of form between tip-growing hyphae and zoospore-spitting sporangia as the pathogen invades the tissues of susceptible host plants, then departs in search of new victims. The airborne dispersal of the sporangia, like the aerial dispatch of any spore, is a passive and deliberately wasteful mechanism. Without a suitable air current eddying

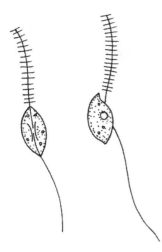

Fig. 7.4 Zoospores produced by oomycete water molds. From N. P. Money, *Mr. Bloomfield's Orchard: The Mysterious World of Mushrooms, Molds, and Mycologists* (New York: Oxford University Press, 2002), with permission.

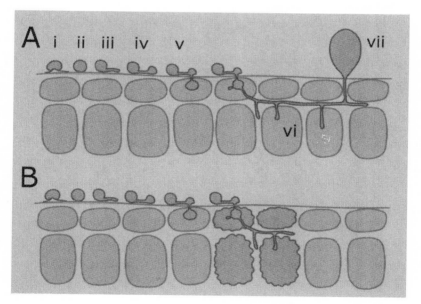

Fig. 7.5 Infection cycle of *Phytophthora infestans*. (A) Infection of a susceptible plant: (i) zoospore settles on leaf surface; (ii) zoospore transformed into cyst; (iii–iv) cyst germinates and forms appressorium; (v) appressorium penetrates epidermal cell of plant; (vi) growth of pathogen hyphae between cells and formation of feeding haustoria within plant cells, and (vii) exit from plant to form sporangium. (B) Stalled infection of a resistant plant. Pathogen penetrates leaf but is confined by hypersensitive death of group of host cells (shown as cells with wavy outline). From S. Kamoun and C. D. Smart, *Plant Disease* 89, 692–699 (2005), with permission.

around a susceptible plant, the sporangium is lost. The loss of most of the sporangia is a necessary gamble on the part of the pathogen that ensures that a few of them strike their targets. Zoospores discharged from this minority of successful sporangia make best use of their rare opportunity and act in a manner that can be likened to self-guided missiles.[6] They swim through the soil or on the plant surface "sniffing" their surroundings, responding to the chemical signatures of their hosts and to interference in the weak electrical fields around plants that may betray sites for invasion.[7] The "olfactory" perception of the zoospore is achieved by the chemical sensitivity of its cell membrane that bristles with receptor proteins. Zoospores are tiny—about 100 of them can fit in a space of one millimeter—but each one carries a hefty cargo of chromosomes in its nucleus that encode around 22,500 genes.[8] This is a lot of genes for a microbe and is not far from the number in humans. With this much information in its cells, why, one might ask, isn't *Phytophthora* good at making anything but potato soup, while humans build skyscrapers? One reason is that lots of cells arranged in a brain can do far more than lots of cells swimming around on their own. How smart, after all, is a human sperm cell?

Like many other pathogens of plants, *Phytophthora* reaches into the juicy cells of its host with a feeding apparatus called the haustorium that looks a bit like a hand. The stubby fingers of the haustoria fit into gloves formed by the plasma membrane of the plant cell. The coffee rust fungus produces a similar structure, which I described in chapter 3, referring to the placentalike connection between the fungus and the plant cell. After forming haustoria, the potato blight pathogen quickly kills and dissolves the potato tissues and absorbs them wholesale into its rapidly growing colonies. The dissolving is done by a cocktail of enzymes that decompose the plant cell walls into simpler sugars. Hyphae absorb the sugars, fueling their ever-extending network of branches. As the colony enlarges, haustoria make the best use of the diminishing population of healthy host cells, and as soon as the cells sicken they are digested. At the moment the appressoria penetrate the leaves of the potato, the plant has a brief opportunity to save itself. If the plant recognizes signals among the haze of molecules emanating from the invading hyphae, it can seal off the point of infection by killing a halo of its own cells to starve the pathogen (fig. 7.5B). I talked about this hypersensitive response in relation to the defenses employed by cacao plants against witches' broom in chapter 4. In susceptible plants, however, the hypersensitive response is suppressed by molecules called effectors, which are secreted by the pathogen.[9]

The ongoing effort to sequence the huge genomes of a number of species of *Phytophthora* is aiding research on the molecular basis of the disease. The implications of this work go beyond the field of plant pathology because of the strange position of water molds in evolutionary history.[10] Sitting far outside the fungal kingdom, the oomycete water molds are believed to have evolved from algae by adopting the absorptive way of life over the photosynthetic habits of their ancestors. Evidence for this is the discovery that the genome of *Phytophthora infestans* harbors genes usually associated with photosynthesis—something that a true fungus, with its shared ancestry with us animals, would never own.[11] The fact that the hyphae of water molds behave a lot like those of mushroom-forming fungi is a spectacular example of the phenomenon of evolutionary convergence. Both groups of microorganisms developed the same trick for eating solid foods. This can be described as a "push and dissolve" process, referring to the mechanical thrusting of the hyphal tips as they secrete digestive enzymes. The mechanical thrusting is achieved by the exertion of a couple of atmospheres of pressure by each microscopic tip.[12] Further similarities between fungi and water molds are apparent in the details of their interactions with plants. For example, I've already mentioned that both groups of plant pests produce haustoria, and both can induce a hypersensitive response. Both produce the infection platforms called appressoria as well. All of this suggests that there are few alternatives to making war against a plant, and the microbes that have done so arrived at remarkably similar answers during the millions of years that they spent perfecting their art.[13] By studying the fundamentals of plant infection, the rules that are common to fungi and water molds, one could be forgiven for hoping that a universal cure for plant diseases would emerge. I put greater odds on winning the lottery without buying a ticket, but the research is very interesting.

The cost of today's research on blight and other plant diseases represents a significant obstacle to progress. Molecular and cell biological experiments can't be done cheaply, and competition for funding is intense. Scientists involved in the genome research on *Phytophthora* have benefitted by pooling resources through international consortia.[14] Reverend Berkeley faced very different challenges. He didn't have a good microscope until Joseph Hooker presented him with one in 1868, which meant that he was limited to using magnifying loupes, or hand-held lenses, in his early investigations on fungi.[15] Even with Hooker's gift, Berkeley found research impossible on dark days when there wasn't enough light to bounce off his microscope mirror. Berkeley suffered the additional distraction from research in the form of his full-time job as a curate in Northamptonshire. These hindrances to the reverend's investigations should humble an inveterate complainer like me, but I'm sure I'll continue to be upset whenever the National Science Foundation fails to fund one of my brilliant grant proposals.

At the time of the potato famine, there were considerable obstacles to solving the cause of the disease. Because plant pathology didn't exist as a scientific discipline, Berkeley had to invent it, and because Benedict Prévost's earlier explanation of cereal smut had been largely ignored, Berkeley had to reinvent the idea that a specific microbe caused a particular plant disease. Berkeley's evidence that the water mold caused the blight rested on the fact that the filamentous colonies always accompanied the disease: no filaments in the leaf tissue meant no disease. Rated against the benchmark of the Tulasne's illustrations, Berkeley's drawings of *Phytophthora* are quite primitive, but he conveyed the impression of the intimacy between the hyphae of the pathogen and the dying potato (fig. 7.2). This was an important message because Berkeley needed to convince his readers that the filaments were infiltrating and destroying the plant. Ernest Large speaks on this point: "If a man could imagine his own plight, with growths of some weird and colorless seaweed issuing from his mouth and nostrils, from roots which were destroying and choking both his digestive system and his lungs, he would have a very crude and fabulous, but perhaps instructive idea of the condition of the potato plant when its leaves were mouldy with *Botrytis* [*Phytophthora*] *infestans.*"[16]

Berkeley's illustrated article on the potato blight was the finest work on plant disease written since Prévost's memoir on the bunt.[17] Unlike Prévost, however, Berkeley was unable to prove that his pathogen caused the disease. Berkeley was a collector of new species, a taxonomist, and his conclusions about the blight rested on observation rather than on experiment.[18] This explains, in part, why the clergyman gets no more than a brief mention in most histories of science, even though he was wrestling with the nature of disease 15 years before Louis Pasteur's classic experiments with swan-necked flasks.[19]

But while Berkeley failed to demonstrate cause and effect, his explanation of the blight was revolutionary in the way that it assaulted superstition. By focusing attention on a microorganism, the clergyman left little wiggle room for spiritual explanations of the potato disease. After Berkeley—appropriating one of my dad's favorite phrases—there was no room for "cloth-eared idiots" in the study of plant diseases.[20] Ironically then, a clergyman succeeded in reducing God to the status of an ambiguous witness. But Berkeley was no atheist, and he massaged faith into his science by concluding, "It is by these instruments, contemptible in the sight of man, that the Almighty is pleased to accomplish his ends."[21]

Those "ends" included the Irish potato famine. The European epidemic was preceded by disease outbreaks in North America that began around Philadelphia and New York in 1843 and spread across the Midwest and into Canada in 1844.[22] The impact on the American potato harvest was severe. Half the crop in Pennsylvania and Delaware was lost to the "new disease" in 1843, but the diverse agriculture of the region ensured that nobody would starve. The pathogen may have arrived in Europe by this time, but it didn't ravage the Irish potatoes until a pattern of cool, rainy weather set in after the summer of 1845. The potato crop had failed in parts of Ireland earlier in the nineteenth century due to "taint," which is a kind of dry rot, and other diseases, but the blight damage was on an entirely different scale.[23] It destroyed 40 percent of the nation's crop in 1845, and wiped out 90 percent of the potatoes when it returned the next year. After a brief respite in 1847, *Phytophthora* dissolved the crop again in 1848.

The blight was epidemic in many parts of Europe, but sociology explains why the Irish suffered so greatly. The population in Ireland increased from 5 million in 1800 to 8 million by the early 1840s, and by then more than 3 million people relied upon the potato crop. It furnished most of the calories for tenant farmers and served as the currency that allowed them to pay their rent. The death toll resulting from starvation and disease was somewhere close to 1 million people, and more than 1 million people emigrated in the famine years. By 1911 continuing emigration had cut the population to half its prefamine level. Ireland was part of the United Kingdom of Great Britain and Ireland in the nineteenth century. The relief effort from Britain was guided by the Poor Law of 1838 and concentrated on assistance to destitute families through an inadequate network of workhouses. The response to the famine was hopeless. Queen Victoria designated March 24, 1847, as a day of prayer for the Irish and donated a fleck of her own fortune for famine relief, but there was no relief concert in Hyde Park.

Some writers, particularly American academics, believe that the death toll during the famine was even higher and reason that this resulted from a policy of mass starvation administered by the British government.[24] Much of this scholarship is pretty shoddy, but I'm going to pause at this juncture without casting my ballot for or against the government. The shameful behavior of the British in Ireland wasn't a big issue in the history classes that I took in school. I have to admit, for example, that I knew nothing about Cromwell's seventeenth-century terror campaign in the Irish countryside until I heard the rock band called The Pogues cursing him in a song called *Young Ned of the Hill*.[25] In light of this ignorance, I won't insult objective experts on the famine with my interpretation, and, instead, recommend Larry Zuckerman's book *The Potato: How the Humble Spud Rescued the Western World* for a recent and illuminating history of the blight.[26] I can speak, however, to the mycological roots of the disaster. A single potato variety called the lumper sustained the poor Irish farmers. It offered zero resistance to *Phytophthora infestans*, and the cool, wet weather of the famine years was perfect for its swimming spores. Utter reliance on a monoculture is, as I may have mentioned, a surefire recipe for disaster.

Berkeley's "solution" to the blight had no practical implications for those afflicted by the epidemic. Anton de Bary turned his attention to the potato fungus in 1860 following his work on the rust life cycle, and soon succeeded in infecting healthy potato plants with sporangia taken from diseased leaves.[27] He was the first to watch the swimming spores of *Phytophthora* emerge from their sporangia and penetrate leaves. When a new generation of sporangia emerged from the leaves of the deliberately infected plants, De Bary knew that he had the proof of causality that eluded Berkeley. But even with this clarification of the infection cycle, there was no cure for the disease. Part of a crop could be protected by harvesting the tubers at the first sign of leaf damage, but there was no way of knowing if a tuber was rotting if it had been infected directly by spores swimming through the damp soil.[28] Therapy for potato blight awaited the discovery of the Bordeaux mixture by a French investigator, Pierre Marie Alexis Millardet, in the 1880s. The story of the discovery of this powerful fungicide has become very blurred in the last 120 years, and the record needs to be set straight.

Millardet had been trained by De Bary and became a professor at the French University of Strasbourg. He served as a medic in the Franco-Prussian War and then returned to academia as a professor in Nancy and later in Bordeaux.[29] In Millardet's telling of the story, he was strolling through the Saint-Julien vineyards in Médoc when he noticed a bluish-white deposit on some of the vines.[30] He learned that the vintners sprayed the vines along pathways with a dye to discourage pilfering of the grapes. The deterrent was verdigris—copper sulfate mixed with lime. Millardet observed that the treated plants appeared to be healthier than untreated ones. The ramble through the wine country led the French scientist to study the efficacy of the copper treatment against downy mildew that attacked the vines.[31] Subsequent experiments led him to the optimal combination of lime and copper sulfate that has since proven useful against many fungal diseases, including potato blight. This is the story that all plant pathologists are weaned upon. I don't believe a word of it.

Eighty years before Millardet, Bénédict Prévost figured out that wheat seeds treated with copper sulfate mixed with lime were protected from smutting. Prévost's investigations were conducted in Montauban and were founded on the longstanding use of lime, and fortuitous addition of copper, as a seed treatment by local farmers. Is it likely that farmers in Médoc, whose southern reach is 200 kilometers from Montauban, happened upon exactly the same chemical mix as a means to discourage grape pilfering? Even if Millardet was telling the truth about the reason that they were using this crude version of the Bordeaux mixture long before his arrival, it seems strange that the farmers hadn't noticed that the treatment also discouraged mildew.

In his memoir of the discovery, Millardet mentions a rumor concerning copper sulfate treatment . He says that this rumor encouraged him to continue his experiments, despite early ambiguous results, and to share his findings with the regional agricultural society. It seems likely, at least to me, that this "rumor" concerned the established use of copper and lime by French farmers. They may have used this mixture as a preventive against all kinds of fungal diseases—whether they steeped their seeds in it to prevent smut, as Prévost recommended, or sprayed it on leaves to escape mildew. In all probability, Millardet didn't invent anything from scratch the way plant pathologists tell it. But he did figure out how best to use copper sulfate as a fungicidal spray. Before Millardet's systematic inquiry, the farmers had little idea about how much to apply to their vines nor when to apply it. He was like the majority of successful scientists: someone who was good at his job, deserved a paycheck, and should have been lucky to get a footnote in the history books. Instead, there is a monument to Millardet in Bordeaux, topped by his bust, with a naked maiden climbing up the monument offering him a bunch of grapes. Hardly seems fair, does it? This is, however, a minority opinion since many plant pathologists refer to "The Millardetian Period" in their history, from 1883 to 1906.[32]

Bordeaux mixture is a preventive fungicide, distinguished from a systemic fungicide that permeates the tissues of a diseased plant. Millardet understood that Bordeaux mixture worked against grape mildew when its spores germinated on the grape leaves and berries, attacking the pathogen when it was most vulnerable. The same insight about vulnerability occurred to Marshall Ward in Ceylon as he witnessed the disintegration of the coffee plantations. A mixture of lime and sulfur, rather than copper, had been recommended by Berkeley and others to combat the coffee rust, but the epidemic had gone too far for this treatment to work (chapter 3). Copper and sulfur work in a similar fashion by interfering with critical enzymes that meet the energy needs of the fungus.[33] Their action is akin to severing the cables of a car battery. Insights about the mode of action of fungicides usually come after a particular chemical emerges as a the most promising compound among a list of candidates, but there is increasing interest in designer fungicides aimed at particular biochemical pathways and steps in a pathogen's life cycle.

Recognition of the effectiveness of copper against fungi came too late for the Irish potato crop, but the Bordeaux mixture has since proven its utility against potato blight for more than a century. It continues to be used as a preventive spray, but its toxicity to earthworms and other soil inhabitants, and the resulting reduction in soil fertility, is one reason that modern synthetic fungicides have replaced this simple blend of chemicals. The list of fungicides used against the blight includes protectants such as mancozeb, translaminar sprays that penetrate the leaf tissues but don't move around the whole plant, and the systemic compound metalaxyl that was launched as a wonder drug against potato blight in the 1970s. Metalaxyl is one of a handful of chemicals that is selective against *Phytophthora* and its relatives.[34] It acts by inhibiting the water-mold version of the polymerase enzyme that strings together RNA molecules that serve as templates for protein synthesis. There is, however, a drawback associated with fungicides that take aim at such specific targets. Just as bacteria circumvent a novel antibiotic by evolving around it in a few years, so the fungi also change, enabling them to disregard the once-lethal chemical agent and to attack their prey with renewed vigor.

In addition to potatoes, *Phytophthora infestans* infects tomato fruits, which isn't surprising because most pathogenic Phytophthoras molest a range of host plants, and potatoes and tomatoes belong to the same family, the Solanaceae. Both crops originated in South America, which is also where the water mold seems to have been born. Disease-causing fungi and their hosts usually grow up together, though the pathogens can extend their choice of prey species whenever they migrate and find an unwitting source of calories. Knowing that the potato came from the Andes, nineteenth-century investigators guessed that its pathogen had come from the same place.[35] Mexico is another potential home for the water mold, but to explore the issue of the origin of the pathogen a little further, we need to think about its sexuality.

Water mold eggs are produced in the following fashion. Male organs called antheridia grow toward female organs called oogonia that contain unfertilized eggs. When these come together, so to speak, the male clamps onto the surface of the oogonium, pierces its wall, and injects sperm nuclei into the eggs.[36] After fertilization, the eggs develop thick walls and are called oospores. This sexual part of the life cycle was established for other water molds in the nineteenth century, but nobody could find the eggs of the potato blight pathogen. British mycologist Worthington Smith thought that he had discovered them in 1875 and illustrated their development in a controversial paper in *Nature*.[37] Having looked for them himself, without success, Anton de Bary was immediately suspicious of this breakthrough. When De Bary examined Smith's microscope slides, he concluded that Smith had illustrated the eggs of another aquatic fungus that fed on the dead and dying potato leaves and tubers in the wake of *Phytophthora infestans*.[38] Despite De Bary's stellar reputation, a retraction was never published in *Nature*, and most investigators continued to believe in Worthington Smith's discovery until the real eggs were found in 1910 by the American plant pathologist George Clinton.[39] Rather than hunting through drops of the rotten potato soup left by the blight, Clinton studied cultures of the pathogen. He found the male and female sex organs—the antheridia and oogonia—and could induce fertilization, but, left alone, the cultured water mold was very "unenthusiastic" and verged upon sterility. When it was paired with a different species of *Phytophthora*, however, lots of hybrid eggs developed. Evidently, the potato pathogen eschewed its own kind in favor of mating with other species. George Pethybridge, who worked for the Irish Department of Agriculture, discovered another unusual feature of the sexual behavior of the potato fungus. In place of the conventional commerce

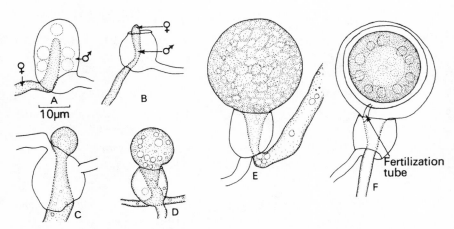

Fig. 7.6 The sexual reproductive process characteristic of the potato blight pathogen. (A, B) The female oogonial branch penetrates the male antheridium, and (C, D, E) its tip swells to form the oogonium or egg sac. (F) The male branch forms a fertilization tube that penetrates the oogonium. Once fertilized, the oogonium contains a single oospore. The drawings were made from *Phytophthora erythroseptica*, which is a relative of the potato blight pathogen. From J. Webster, *Introduction to Fungi* (Cambridge, UK: Cambridge University Press, 1980), with permission.

between the sexes, fertilization was preceded by the penetration of the sperm-laden male branches by the microscopic female sex organs of the potato fungus (fig. 7.6).[40]

The reason that the potato pathogen was such a poor sire was solved when it was discovered that it had two mating types, designated A1 and A2. When A1 meets A2, genes are recombined, and offspring with new characteristics are born. A single mating type will generate male and female organs when coerced to do so on a culture dish, but, as Clinton found, fertility is exceedingly limited. In all likelihood, *Phytophthora infestans* doesn't form oospores very often, instead reproducing asexually through the formation of zoospores and spreading in the form of genetically distinctive lineages in different parts of the world. A handful of different pedigrees have been identified, labeled Ia, Ib, IIa, and IIb, and are based on diagnostic DNA sequences housed in their mitochondria.[41] Each of these strains exists in two mating types: Ia/A1, Ia/A2, Ib/A1, and so on.

This brings us back to the idea that the pathogen evolved in Mexico. This theory of origin became popular when both mating types of the water mold were discovered in the central part of the country, accompanied by lots of genetic diversity—precisely what one would anticipate of the pathogen's birthplace.[42] Everywhere else that the *Phytophthora* had been studied, investigators had found celibate A1 strains of the Ib group. If the Ib pathogens that were rotting potatoes in Europe in the twentieth century were offspring of the strain that caused the famine, this would suggest that a single strain had escaped from Mexico in the 1840s and caused all the suffering.

But just when scientists thought that everything was fitting neatly into place, the A2 mating type was found in Switzerland in 1984, and then it popped up just about everywhere else that the water mold was collected. The arrival of A1's ostensibly long-lost mate meant that *Phytophthora* could be mating a lot more than anyone had imagined possible. Plant pathologists found this disturbing because the restoration of its ancient sexuality made it more likely that the species would develop fungicide resistance and become even more aggressive. This was, unfortunately, what potato farmers were already beginning to report. Outbreaks of late blight of potato and tomato assumed epidemic status in the United States and Canada in the 1990s.[43] Fields of plants decayed despite spraying with metalaxyl, and farmers were photographed atop piles of rotting tubers. Fears were raised about the vulnerability of crops in Russia, where there is limited access to fungicides and disease-resistant seed varieties.[44] Somehow, 150 years after the famine, the genetic diversity of the pathogen that had been migrating around the potato-growing regions of the world had been boosted by the introduction of new strains from Mexico and South America. Today, according to potato blight expert Howard Judelson, "The worldwide cost of the potato disease alone exceeds $5 billion per year, including $1 billion spent on fungicides. This is enough to purchase potatoes to fulfill the caloric needs of the entire world for 2.7 days, based on 2200 Kcal/day and current U.S. potato prices!"[45]

In the last few years, the picture of the blight's origin has been illuminated by the results of molecular detective work of Jean Ristaino at North Carolina State University. Ristaino began her research by reviewing samples of diseased potato plants that had been collected in the nineteenth century and archived in herbarium collections. The first interesting discovery came when she looked at specimens collected in Britain the 1870s and found eggs that looked like those of *Phytophthora infestans*.[46] The samples of infected stems and tubers had been deposited at Kew by a botanist named

Charles Plowright who had noted the presence of the oospores on his specimen labels. De Bary, assisted by two research assistants, had pored over infected potatoes from the same time period and never found the eggs. Bad luck or poor judgment must have intervened: Had they missed the oospores, or had they seen them, but thought they were something else? Worthington Smith believed that the issue had become politicized to the point that nobody dared question De Bary's original dismissal of the eggs. If the eggs that Ristaino rediscovered in the herbarium samples were those of the pathogen, then perhaps both mating types had traveled around Europe in the nineteenth century.

Ristaino approached the issue of the pathogen's genetic identity by extracting DNA from herbarium samples, including some collected at the time of the famine. She succeeded in amplifying DNA from specimens that had belonged to Berkeley's herbarium and found that the microbe responsible for the potato famine belonged to the Ia lineage, not the Ib type that was responsible for the recent disease outbreaks.[47] Because the Ia type is found in Mexico and South America today, the molecular research doesn't pinpoint the origin of the microbe responsible for the Irish famine. South America remains the best bet, however, because potatoes were not exported from Mexico in the 1840s. How then did the microscopic water mold travel thousands of miles to Europe?

Let's consider a few possibilities. It isn't likely that the infectious zoospores swam from the Andes to Europe, but this idea does serve as an interesting (though futile) thought experiment. Traveling at a speed of 0.1 millimeters per second, a *Phytophthora* zoospore would have to swim for more than 3,000 years to cover the distance of 10,000 kilometers. To sustain this feat of endurance the cell would consume at least 3 million times its own "body" weight in food.[48] An additional concern for the water mold is the toxic salinity of seawater. What about the airborne sporangia? A meteorologist in need of a publication might be able to stitch together some wind patterns to waft sporangia all the way from Lima to Limerick, but I doubt that anyone would take it seriously. The escape of the pathogen was probably much simpler: it is most likely to have come on ships laden with bird feces and tubers. Exports of guano from South America to North America and Europe began in 1840, and shipments of the raw material for explosive manufacture were accompanied by potatoes.[49] Some of those potatoes must have been blighted with the Ia strain.

Although these are early days in the molecular exploration of the pathogen's movements, the contemporary prevalence of the Ib strain suggests that the famine pathogen was usurped by fresh strains arriving from Mexico or South America. The virulence of the new immigrants may have allowed them to outstrip the original microbe wherever they encountered one another in a leaf or tuber. It is, after all, a dog-eat-dog world. By today's standards, that Ia strain may have been a weakling, but that didn't mean much when it landed in the waterlogged fields of Ireland. One morning in 1845, a cloud of pioneer sporangia blew across a crop of brilliant green lumpers and stuck to their wet leaves. All epidemics start small. This one led to the deaths of one million people and incited the largest human migration in history. In his poem "At a Potato Digging" (1966), poet Seamus Heaney captures the horror of life for the tenant farmers after the arrival of blight:

The new potato, sound as stone,
putrefied when it had lain
three days in the long clay pit.
Millions rotted along with it.

Mouths tightened in, eyes died hard,
faces chilled to a plucked bird.
In a million wicker huts
beaks of famine snipped at guts.

A people hungering from birth,
grubbing, like plants, in the bitch earth,
were grafted with a great sorrow.
Hope rotted like a marrow.

Stinking potatoes fouled the land,
pits turned pus in filthy mounds:
and where potato diggers are
you still smell the running sore.[50]

Blights, Rusts, and Rots Never Sleep: Forestry and Agriculture, Biological Warfare, and the Global Impact of Fungal Disease

Background

Cooksonia, the earliest land plant, a stringy thing outfitted with spore-filled balls, appears in the fossil record from the Silurian Period, 425 million years ago. Fungi are around at this time, but none of the plant remains bear unequivocal scars of pathogens.[1] Devonian fossils reveal aquatic fungi (chytrids) that were caught in the act of infecting algae 400 million years ago,[2] and the intricately branched cells of mycorrhizal fungi (microbes that are essential to plant health) are entombed in fossilized roots of the same age. The absence of fossils of diseased land plants of this age doesn't mean that the fungi were nicer to plants back then, but more likely reflects the bias of paleobotanists who collect the best looking fossils and discard the compressed remains of rotting tissues. Mycorrhizal fungi were probably with the land plants from the beginning, mining scarce nutrients for their partners in exchange for some solar-generated sugars. Some of these cooperative fungi may have become nasty toward plants, or perhaps the nasty ones were on the scene long before the mycorrhizal species evolved. In any case, I bet that there was an ancient oomycete water mold—*Protophytophthora cooksoniensis*—that chewed on the plants growing in the Silurian mud.

Current evidence dates the evolution of the ascomycete fungi (the group that includes the fungi that cause chestnut blight and Dutch elm disease) to around 300 million years ago, and this was the prelude to the birth of many of the modern groups of plant pathogens. By the time humans came along and invented agriculture, the fungi were already well equipped to dissolve our crops. Our monoculture approach to farming, linked intimately to relentless deforestation, furnished the microbes with titanic feeding grounds of utter genetic uniformity. Rusts and smuts confounded farmers and famished us for millenia, but when the cumbersome mechanical harvesters were dragged across the prairies in the late nineteenth century, they spread the biggest spore clouds in history—at least in *human* history. Then came plant pathology and fungicides, and if we had been more prudent farmers and conservationists and had committed to global birth control, we might have created a twenty-first-century Eden. Fortunately for plant pathologists, and unfortunately for the rest of the biosphere, we failed our planet on a number of counts. In this final chapter, I explore instances of the continuing advance of the fungi and consider the status of the ongoing competition between humans and fungi for control of the earth.

Fungal Threats to Forest Resources

White pine blister rust, caused by *Cronartium ribicola*, kills white pines, also known as five-needled pines (fig. 8.1). The disease has been around as long as chestnut blight, but unlike the chestnut pathogen, the pine fungus has latched upon an entire family of vulnerable hosts in the western United States.[3] The first symptoms of blister rust are yellow or reddish spots on needles. These appear close to the points of entry of the fungus. During the next couple of years, the plant's attempts to throw off the infection result in misshapen growths on the branches. Eventually, these cankers strangle the tree limbs, resulting in leaf fall and destruction of the most active parts of the plant. The life cycle of this rust is similar to *Puccinia graminis*, the only pertinent difference being the hosts. *Cronartium*'s alternate hosts are currant or gooseberry bushes (*Ribes* species). The rusty uredospores, the teliospores, and the basidiospores are produced on the currant leaves; white pine is infected by the basidiospores, which travel from sore to sore on the tree, and once sexual reproduction is complete, the fungus broadcasts its currant-infecting aeciospores into the air. An infection cannot be

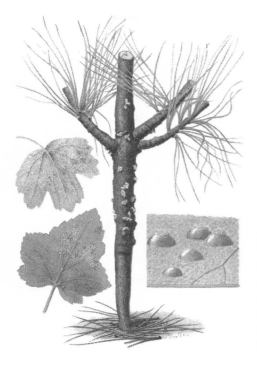

Fig. 8.1 Symptoms of white pine blister rust caused by *Cronartium ribicola*. From W. V. Benedict, *History of White Pine Blister Rust Control—A Personal Account*, USDA Forest Service FS-355 (Washington, DC: U. S. Government Printing Office, 1981).

transmitted from pine to pine, but this hasn't done much to obstruct the disease.

White pine blister rust was introduced from Germany into the United States at the end of the nineteenth century; it was discovered by Fred Stewart at the New York Agricultural Station in Geneva in 1906.[4] Again, the historical confluence of these stories of disease introductions is striking: chestnut blight was discovered in 1904 (and was also studied by Stewart), the European Dutch elm epidemic developed in the next decade, and so on. In the case of pine blister rust, human culpability is very clear. The disease was well known in Russia in the 1850s and spread westward across Europe. The threat to American forests was obvious, but persistent warnings from scientists, including Carl Freiherr von Tubeuf, who was a pioneer in the biological control of plant diseases, didn't prevent multiple imports of white pine seedlings bearing the rust, including a West Coast introduction to Vancouver Island in 1910.

This transatlantic movement of the disease is part of a complicated story of international trade. Neither of the rust's preferred hosts grew in Europe before the eighteenth century. Currants and gooseberries were planted earlier than this, but it wasn't until they became fashionable and were cultivated widely that the fruity half of the rust's life cycle could be richly supported. Meanwhile, transatlantic plant commerce provided the fungus with its other host. With a dwindling supply of native timber, Europeans began importing white pine seedlings from North America in an attempt to refurbish their forests in the 1700s. White pines were planted all over Europe, especially in Germany and the Baltic States, and as far as western Russia. The fungus probably arrived on pines or currants collected for botanical gardens between 1750 and 1850. By this time, rust paradise existed just beyond the brick walls of the gardens in the form of countless susceptible fruit bushes and dense plantations of white pines. The welcome could not have been warmer.

Back in America, vast swathes of the pine forests had been cut in the eastern United States. The market for white pine had prospered since 1605, when Captain George Weymouth of the British Royal Navy had recognized that big white pines could be turned into ship masts. The smaller ones were perfect for just about everything else that humans do with wood, and demand for white pine provided pre-revolutionary America with one of its earliest industries. Sawmills buzzed through seemingly endless streams of logs felled in the virgin forests. White pine exports drove one side of the triangular trade that brought slaves to the Caribbean: the American lumber was carried to West Africa in ships made from planks of white pine; African slaves were taken to the West Indies, and sugar and rum were transported to New England. The pines continued to fall through the Revolution and the end of slavery, and after two centuries of deforestation, replanting seemed like a splendid idea.

When American nurseries couldn't meet the domestic demand for white pine seedlings, millions of seeds were supplied to German growers. Many of the white pine seedlings that came back had acquired a dusting of rust spores in Europe. To be sure that you are following this, let's recap the journey of the pine and its rust: (1) pine is introduced from the United States to Europe and Russia in the eighteenth century; (2) rust spreads through forests of the introduced white pine in the nineteenth century; (3) pine seedlings, now rusted, are reintroduced to North America, where an epidemic of blister rust explodes in the twentieth century. The original, pre-European homeland of *Cronartium* isn't known for sure, but in common with our contemporary picture of chestnut blight and Dutch elm disease, evidence points to an Asian origin for blister rust.[5]

White pine blister rust became a gargantuan problem in the northeastern United States. To combat the epidemic, pines and currants were eradicated in broad rings around the sites of the initial outbreaks in 1909, an Office of Blister Rust Control was established, and conferences were organized—all mirroring the government response to chestnut blight.[6] Later, a quarantine ordered the destruction of cultivated currants and gooseberries wherever they grew close to white pines, and this was followed by a massive national effort to eradicate wild *Ribes* plants.

Ronald Thaxter, a plant pathologist at the Connecticut Agricultural Experiment Station, refused to allow government workers to exterminate currant bushes in the garden of his summer home in Maine. At what became known as The Battle of Kittery Point, Thaxter met the agents at his gate with his shotgun. Given his job title, Thaxter's stand was peculiar, but it typified his schizophrenic view of plant pathology. He called plant pathology "squirt gun botany" in reference to fungicide application, which deepens the irony because he was responsible for some of the pioneering research on the use of Bordeaux mixture.[7]

Thaxter participated in a squabble between plant pathologists and mycologists that has existed for decades. The unofficial dispute concerns the relative superiority of the two scientific disciplines. Plant pathology is a broader subject than mycology because in addition to the fungi it encompasses the study of insects, nematodes, bacteria, and viruses that attack plants. The fungal part of the field is thoroughly intertwined with mycology, but friction has developed by the perceived competition between basic and applied research. While plant pathology has always been concerned with practical issues, mycology began as a pure celebration of biological diversity. Today, however, this distinction is meaningless. Like all scientists, contemporary mycologists are also forced to justify the practical applications of their work. A mycologist studying how hyphae grow across a Petri dish has the potential to influence the way that an expert on chestnut blight understands disease development. Similarly, a clever insight about the way that a particular fungicide works may influence a far broader understanding of the way that all fungi grow. Plant pathologists obtain doctoral degrees in plant pathology, whereas many mycologists are born in botany or biology departments, but there are so few fungal biologists of any provenance that we should all kiss and make up.[8]

Thaxter's cynicism toward the eradication of *Ribes* was partly justified. The program proved effective at protecting pockets of pine trees, but overall the rust was unhindered in its sweep through the evergreen forests of the United States and Canada. By the 1930s, white pines in New England had suffered 40 percent mortality, and mature trees vanished from many parts of the Pacific Northwest. Regeneration was impossible because seedlings were exceedingly vulnerable to the rust. A century after its introduction, *Cronartium* is ranked among the greatest threats to American forests. In the last 50 years, however, the incidence of the disease has shown a marked decline in the Northeast. The reason for this is unclear but could be explained by the natural selection of resistant trees in the early years of the epidemic.[9] This is, perhaps, how diseases are supposed to work. The coevolution of pathogenic fungi and their hosts usually prevents the extinction of either participant when there is plenty of genetic diversity among the hosts, but mutual annihilation becomes more likely when a pathogen encounters a monoculture. As I have illustrated in this book, there is a profound difference between the outcome of disease outbreaks in natural ecosystems versus epidemics among cultivated species. Immunity to a new disease is less likely to exist, or to develop, among a population of genetically similar crop plants. This is also true of trees like chestnuts and elms that exist as cryptic monocultures after passage through a genetic bottleneck in their history.[10]

Today's blister rust frontline is in the western United States and Canada where most species of white pine grow. One of the hosts is western white pine, *Pinus monticola*, whose timber is vital to the economy of this region. Whitebark pine, *Pinus albicaulis*, is another victim: this species offers the lowest level of resistance to the disease among all the pines. Whitebarks have vanished from areas where the disease has been established for decades, and more than half of these trees have been killed in some parts of the Greater Yellowstone Area. This has serious ramifications for the rest of the wildlife in Yellowstone because whitebark forests cover a quarter of the landscape. Whitebark pine is a "keystone" or "umbrella" species that is essential to the health and survival of the entire ecosystem.[11] The tree thrives at high elevations, and in these remote locations, where human interference is limited, its fatty seeds are an important part of the diet of grizzly bears.[12] Indeed, in some years, the seeds account for the majority of the calories consumed by the animals. With this information in hand, you don't need an advanced degree to surmise that the rust is a severe problem for *Ursus arctos*. Studies have found that bears forage in lowland areas closer to human habitation when there are shortages in pine seeds, which isn't good for grizzlies or for us. Naturalist John Muir once expressed the interconnected nature of living things when he said, "When we try to pick out anything by itself, we find it hitched to everything else in the universe."[13] The connection between blister rust in the mountains and a hungry bear crashing through your kitchen door exemplifies this to a tee.

The rust poses other problems. Bristlecone pines and sugar pines are susceptible to the disease. Bristlecone pines, *Pinus longaeva*, are among the oldest living organisms, with a specimen in the White Mountains of California named "Methuselah" approaching the end of its fifth millennium.[14] Some years ago, experiments showed that bristlecones could be infected by *Cronartium*, and the first natural case of the disease was discovered in 2003 in the Great Sand Dunes National Monument in Colorado.[15] The afflicted tree was surrounded by heavily rusted limber pines (*Pinus flexilis*). Sugar pine, *Pinus lambertiana*, is the tallest pine in North America and has been heavily damaged in the Central Cascades and Sierra Nevada.

Given enough time, the virulence of the disease among the western species of white pine is likely to be dampened according to the model in the northeastern forests. Natural selection will favor those trees with enhanced resistance, and as the genetic map of the forests is redrawn, the rust fungus will be forced into a holding pattern of infrequent infection—until, of course, it mutates and bursts through another virulent phase in its evolutionary history. But for those of us sentimental enough to wish for the survival of individual trees like Methuselah, the mere existence of the rust is dreadful news.

White pine blister rust is a big problem for American forestry, but it has been eclipsed recently by an even bigger story in plant pathology. Sudden oak death, or SOD, is caused by a new species of pathogen with an alarmingly broad host range. The disease is incurable, and it threatens the American landscape unlike anything since chestnut blight. Problems among oaks were recognized on the West Coast in the mid-1990s. The trunks of species native to this region developed enormous patches of dead tissue: cankers that bled black or reddish sap through the thick bark (fig. 8.2). The leaves turned brown within a few weeks and, eventually, the entire crown of the tree perished. Symptoms of plant diseases don't come any more dramatic. Journalists seeking an alarming headline would have been happier,

Fig. 8.2 Symptoms of sudden oak death revealed by removing bark from a coast live oak. Patches of decaying phloem tissue and underlying cambium are evident in the exposed area. Photograph courtesy of David Rizzo.

of course, if mature trees began exploding along California highways, showering commuters with shards of bark. But even in the absence of human casualties, plenty had been written about SOD by the time its cause was identified by plant pathologists David Rizzo, Matteo Garbelotto, and colleagues in 2002.[16]

The pathogen that caused SOD was first isolated from trees by slicing into the cankers and transferring samples of discolored tissue to agar medium in Petri dishes. Colorless hyphae grew out from the chunks of tissue into the agar. To test whether this was the cause of the disease, the investigators inoculated tree seedlings with plugs from the cultures. They also infected young saplings and some mature trees. In all cases, plants inoculated with the unnamed microbe developed symptoms of the disease, and controls inoculated with clean agar plugs remained healthy. Finally, Rizzo and colleagues isolated the same organism from the experimental infections. It looked like a species of *Phytophthora*, a relative of the potato blight pathogen.

These experiments followed the standard protocol for pathogen identification—the same strategy used, for example, by Marie Schwartz in the 1920s when she figured out the cause of Dutch elm disease (chapter 2). But the SOD investigators also used modern molecular methods to determine the identity of their *Phytophthora*. Molecular methods were necessary because it is exceedingly difficult to distinguish between the 60 or more species of this microbial genus using a microscope. They all look similar growing on agar, though experts can separate some of them based on the microscopic details of their zoospore-producing sporangia, sexual organs (antheridia and oogonia), their eggs (oospores), and thick-walled spores called chlamydospores. Briefly, DNA is isolated from the cultured cells (or from infected tissues), and particular regions of the pathogen's genome are amplified (copied) using the polymerase chain reaction, or PCR. A bench-top instrument called a thermocycler is used to make millions, sometimes billions, of copies of the DNA sequence in the course of a few hours. Once this is done, the sequence can be read using an automatic sequencer, and the string of adenines, thymidines, guanines, and cytosines displayed on a computer monitor can be compared with the sequences from the organism's relatives. In the initial study the investigators compared the DNA from the SOD pathogen with 44 *Phytophthora* sequences that were available in GenBank, an online database. The simplicity of this cookbook identification of microbes is horrifying for the dwindling number of scientists with expertise in classical microscopic identification, but the power of the modern technology is undeniable.

The appearance of the SOD pathogen didn't match any of the recorded species of *Phytophthora*, nor did its DNA sequence. Unbeknownst to the American investigators, European pathologists had been studying a new species of *Phytophthora* that had been killing rhododendron and viburnum bushes in plant nurseries and gardens in Germany and the Netherlands. They named it *Phytophthora ramorum*.[17] When Rizzo and colleagues were alerted to this work, they compared the microscopic appearance of *P. ramorum*, and its DNA, with the SOD pathogen and found a perfect match.[18] Two populations of the same *Phytophthora* species separated by 9,000 kilometers were killing oak trees and ornamental bushes.

By the time the SOD pathogen was identified, forests in northern California showed signs of widespread stress. North and south of San Francisco the *Phytophthora* was killing a variety of species including tanoaks (*Lithocarpus densiflorus*), coast live oaks (*Quercus agrifolia*), and California black oaks (*Quercus kelloggii*), marring some of the prettiest countryside in North America. A survey published in 2001 showed that half of the trees in some stands were already infected.[19] Investigators were further dismayed by the discovery that the pathogen was attacking other tree species, including big leaf maples, poison oak, and filberts. *Phytophthora ramorum* didn't limit itself to flowering plants, but also showed up on conifers: grand fir, Douglas fir, yew, and, most alarming of all, the coastal redwood, *Sequoia sempervirens*. The infected redwoods showed dead needles rather than any cankers on their giant stems, but the growing host range was a chilling discovery.[20]

The distribution of the pathogen on nursery plants in Europe suggested that shrubs in garden centers could represent a reservoir of infection in California. This hunch proved sound when the *Phytophthora* was isolated from rhododendrons growing in containers in a Santa Cruz nursery. The California Department of Food and Agriculture responded with an emergency regulation requiring permits for the movement of any of the host species from the affected areas, and Oregon and Canada implemented quarantines against Californian plants. These measures were too late, however, to prevent the appearance of the disease in Oregon. In July 2001, the SOD pathogen was discovered on diseased tanoaks and shrubs in a 40-acre area in Curry County on the California border. This was very troubling because tanoaks grow alongside Douglas fir, Douglas fir is a host for *Phytophthora ramorum*, and Douglas fir is the mainstay of the timber industry's multibillion dollar harvest in this region. The USDA Forest Service and the Oregon Department of Forestry moved swiftly, cutting and burning every infected tree and shrub and every potential host in the affected area.[21] Oregon's first outbreak of SOD was extinguished.

By the time the pathogen had been properly identified, it was too widespread in California for the cut-and-burn remedy, and federal and state agencies invested in a strategy of monitoring and containment rather than eradication. By summer 2005, tens of thousands of oaks had been lost to SOD in a 600-square kilometer heavily forested area that spans 14 counties.[22] The way that *P. ramorum* spreads between plants is not certain but probably involves the short-range locomotion of swimming zoospores through soil plus long-range distribution of the sporangia containing these spores. Sporangia could be spread in clumps of soil on hikers' boots, car tires, and by animals, and individual sporangia might become airborne (especially from infected leaves), like those of the potato blight pathogen.

The future impact of SOD is unclear. The loss of oaks and other trees is destroying vital habitat for wildlife, and further damage will increase water runoff and soil erosion. By killing trees the disease could also worsen the already critical risk of widespread forest fires in some of the affected areas. The danger of forest fire is ironic, because foresters have found that SOD is worst in areas in which fires have been suppressed.[23] Trees may suffer a progressive weakening of their defenses in old, unburned stands, suggesting that a program of controlled burns may be a good approach for containing the spread of the disease. The bleakest forecasters maintain that SOD may decimate western forests. The Mediterranean climate of northern California may be perfect for the pathogen: the nidus of the epidemic receives plenty of winter rainfall, and coastal fogs bathe the trees in moisture throughout the year. The Pacific Northwest is one of the planet's mycological wonderlands, but a climate that is good for mushrooms may also favor destructive microbes. Drier inland areas may obtain some protection from the disease, though there isn't much comfort in this for foresters in Oregon, Washington State, and British Columbia.

Since the discovery of the disease, researchers have been concerned by the susceptibility of the mixed deciduous forests of the eastern states. *Phytophthora ramorum* attacks eastern oaks and maples under greenhouse conditions, but the way that the pathogen would operate in nature is unknown. Fears on this count were heightened in 2004 when a red oak on Long Island tested positive for the *Phytophthora*.[24] Careful examination of putative cases of SOD are important because the disease may be confused with other problems including oak wilt, caused by the fungus *Ceratocystis fagacearum*,[25] and oak decline, which is a slow-acting problem in mature trees caused by drought, insect damage, and root infections by species of *Armillaria*, or honey mushroom.

If SOD does spread in the eastern United States, infected nursery plants and their human handlers will probably be responsible for the catastrophe. The USDA Animal and Plant Health Inspection Service (APHIS) has found that Californian nurseries where *Phytophthora ramorum* has been isolated have shipped thousands of plants around the country since the implementation of the quarantines. One facility in Los Angeles County dispatched potentially infected camellias to 783 garden centers in 39 states.[26] Surveillance by APHIS identified *Phytophthora ramorum* at 121 sites outside California by the end of 2004, mostly on nursery plants.[27]

Meanwhile, *Phytophthora ramorum* has been moving around Europe, causing similar alarm. The majority of the European infections have been identified on rhododendrons and other ornamental plants, but the pathogen has been found on a few trees in the United Kingdom and the Netherlands. In Cornwall, bleeding stem infections were discovered on European beech and horse chestnut trees growing close to contaminated rhododendrons. The native British oaks, *Quercus robur* and *Quercus petraea*, appear to be more resistant to SOD than the dominant red oak of the eastern United States, and foresters are more concerned about the immediate effects of *Phytophthora ramorum* on beech in the United Kingdom.[28]

There is an important genetic wrinkle to the story of SOD. In the previous chapter I explained that distinct mating types of the potato blight pathogen used to be confined to Europe and Mexico. The same type of isolation is true of the SOD *Phytophthora*. Currently, A1 is based in Europe and A2 in California. The origin of the pathogen is a mystery, but the fact that the mating types are so widely separated is consistent with the idea that separate introductions of the microbe occurred on the two continents.[29] Alternatively, *Phytophthora ramorum* may have been around for a long time in both locations and taken divergent evolutionary pathways. The separation of the mating types means that there is no opportunity for sexual reproduction, and the resulting lack of gene flow may act as a brake on the virulence of the SOD pathogen. Crossing of the A1 and A2 mating types of the potato blight pathogen signaled a new phase in the organism's virulence (chapter 7), which raises legitimate concern about the possibility of a future hybridization between the American and European strains of the SOD pathogen. Given the near sterility of *Phytophthora ramorum*, however, such a union, if it ever occurs, may be of limited significance. *Phytophthora* species are the most exuberant swingers of the microbial world, forming male and female sex organs on the same colony, and mating with other species at every opportunity. An A1 of *Phytophthora infestans*, for example, will pair with the A2 of other species of *Phytophthora* to create "interspecific hybrids." This behavior has tremendous potential for the evolution of new species of pathogenic water molds.[30] *Phytophthora ramorum* is the wallflower of the genus: it can be coerced to produce a few distorted sex organs when it is paired with a foreign species, but it rarely consummates these unions. The species doesn't perform much better when its American and European mating types are brought together, which explains why concern about a transatlantic marriage is muted.[31]

More troubling, to me at least, is the ecological complexity of SOD. A couple of other Phytophthoras—*P. nemorosa* and *P. pseudosyringae*—lurk in the cankers of sickened oaks in California. These are less aggressive than *Phytophthora ramorum*, but they offer a glimpse of the chaotic environment from which a novel and spectacularly virulent disease emerges. Sickened by water molds, the oaks develop other secondary infections by beetles and ascomycete fungi. Another *Phytophthora*, known only as taxon C at the moment, is causing bleeding cankers on European beech and leaf and twig blight on rhododendrons in the United Kingdom.[32] Cataloging the sweep of hosts that interest this alien beast is a work in progress.

ॐ

Australian foresters have decades of experience in dealing with a deadly water mold, and, on the whole, the triumph has been on the microbial side. First documented in 1921, a disease swept through the forests of Western Australia, killing the dominant jarrah trees, or *Eucalyptus marginata*, along with 50–75 percent of the other plant species in its path (fig. 8.3).[33] An unidentified microbe killed the indigenous grevilleas, banksias and proteas, the tea trees, acacias, boronias, and varied eucalypts, and left a barren landscape of sedges and rushes in its wake. The disease became known as jarrah dieback and was aptly described as a biological bulldozer. Loss of the natural plants meant loss of the animals. Biologists witnessed the total collapse of the jarrah ecosystem.

Since its discovery in Western Australia, dieback has spread around the coast, causing epidemic disease in Victoria, New South Wales, and the rainforests of Queensland. It also has wrecked thousands of hectares of forest in Tasmania. The water mold has killed prime specimens of enormous eucalyptus trees and threatens critically endangered native flowering plants with extinction. Cone-bearing trees are targets, too. Biologists studying the Wollemi pine, the famous living fossil discovered in New South Wales in 1994, regard *Phytophthora cinnamomi* as a significant threat to its continued survival.[34] A government report published in 2001 concluded that "The disease is in the middle of its epidemic development and is established in a mosaic over millions of hectares."[35]

In contrast to the swift identification of the cause of the contemporaneous epidemics of chestnut blight and Dutch elm in the Northern Hemisphere, the Australian microbe wasn't collared until the 1960s.[36] The delay doesn't speak well for antipodean plant pathology, but the pathogen can lie dormant for

Fig. 8.3 Jarrah forest in Australia decimated by the water mold *Phytophthora cinnamomi*, a superparasite that attacks hundreds of different plant species. This photograph shows dead and dying *Eucalyptus* trees. Photograph courtesy of P. B. Hamm; reproduced with permission from *Selected Plant Pathogenic Lower Fungi*, 1990, APS Slide Set Series, The American Phytopathological Society, St. Paul, MN.

years in soil and gravel and is notoriously difficult to coax out of hiding. *Phytophthora cinnamomi* (named for its original isolation from cinnamon trees in Sumatra) attacks root systems and destroys the food-translocating phloem tissue and revitalizing cambium. Infected plants appear drought stressed and later appear to be dead—which is, indeed, what they soon become. The dieback part of the name refers to the damage to the crowns of the trees resulting from the hidden destruction of the roots. Start to finish it may take *Phytophthora cinnamomi* three years to kill a big tree. One of many puzzles about the disease is that this water mold, equipped with swimming spores, is so effective in the dry soils of the jarrah forests, which are also known as dry sclerophyll forests. This makes sense, however, when the Mediterranean climate of this part of Australia is considered. Although the summers are usually hot and very dry, the forests receive lots of rainfall during the winter months. Like the SOD fungus, *Phytophthora cinnamomi* can cope with dry conditions, but being a water mold it really loves wet weather. Once in a while the region is soaked by torrential summer rainstorms, and these years are invariably followed by ferocious outbreaks of dieback.

Fungicides are only of limited use against dieback. The first thing to consider is the pseudofungal nature of the pathogen, which means that many of the chemicals that kill rusts and smuts are useless against oomycete water molds. Individual plants can be protected with metalaxyl and fosetyl-aluminum, which are systemic anti-oomycete compounds. These chemicals can be applied by drenching the foliage and surrounding soil, or, more effectively by direct injection into the plant. But an entire forest cannot be protected in this fashion. Australian researchers are looking at the widespread application of neutralized phosphonic acid, or phosphite, that seems to be effective at halting the growth of *Phytophthora* inside plant tissues. Phosphite is so cheap that it can be sprayed from aircraft over areas containing key plant species. Because it is rapidly degraded in soils, it poses a limited threat to the purity of the groundwater.[37] Research on phosphite is continuing, but the current strategy for controlling jarrah dieback, like the battle against SOD, is focused on blocking the export of plants and soil from infected areas. Mapping the extent of the epidemic is crucial to these efforts, and this has been aided by high resolution satellite imaging of changes in vegetation.[38]

Like many of the cereal rusts, *Phytophthora cinnamomi* is a super pathogen of global distribution. It rots roots in managed forests and Christmas tree farms; it affects woody nursery plants, and it destroys avocado crops, accounting for hundreds of millions of dollars of losses every year. Both of its mating types are found in Australia, but A2 is far more prevalent. Both lineages also live in Papua New Guinea and South Africa: A1 occurs in forests and heathlands where it doesn't cause disease, and A2 attacks cultivated plants at lower elevations.[39] If you think about Gondwanaland—the supercontinent in which South Africa, Australia, and Papua New Guinea were stuck together with the other Southern Hemisphere land masses—the idea of a longstanding presence of the same microbe in these countries seems logical. But the fact that native Australian plants show zero resistance to the water mold argues against this. Their helplessness is one of multiple signatures of an introduced or invasive pathogen, but the evidence isn't conclusive. Many outbreaks of dieback occur close to roads, suggesting that sporangia are carried on tires. But road construction also affects soil drainage, possibly exacerbating the disease by creating strips of waterlogged soil. Another possibility is that the pathogen is an ancient resident that has been aroused by the enrichment of the naturally nutrient-poor soils of southern Australia by agricultural runoff. Fire suppression might have contributed to the disease, but the evidence is thin. Although forest fires have suppressed dieback in some areas by heating the soil enough to kill the pathogen, in other cases, *Phytophthora cinnamomi* has been one of few microbial survivors after fires and has then proven even more destructive. This should serve as a cautionary tale for those encouraging forest fires for controlling SOD in California.

In Spain and Portugal, *Phytophthora cinnamomi* has been implicated in the decline of cork oak, *Quercus suber*, and another evergreen species called holm oak, *Quercus ilex*.[40] In the United States, the same water mold is a major pathogen of avocado, which is grown in southern California, and also kills a variety of tree species. But with epidemic root rot restricted to Australia, and with *Phytophthora ramorum* hogging the limelight in the Northern Hemisphere, its easy to forget its omnipresent cousin. This may be a mistake. Long before the arrival of chestnut blight, *Phytophthora cinnamomi* wiped out American chestnuts in the foothills of the southern Appalachians.[41] An epidemic of crown dieback was reported among these forest giants in the 1820s.[42] The outbreak followed a period of very wet weather, but the microbe is thought to have been imported much earlier. By the time of the Civil War, the chestnut had vanished from much of its southern range. Decades later, as we saw in the first chapter, the *Cryphonectria* blight would kill the tree everywhere else.

Agriculture and Biological Warfare

Fungi continue to make considerable gains in the agricultural sphere: the list of invasive, emerging, and developing crop diseases is overwhelming, so forgive me if I miss your favorite plant disease, but I'll just mention a couple more of the important ones.

Asian soybean rust, *Phakopsora pachyrhizi*, arrived in the southeastern United States in 2004.[43] The fungus is quite virulent, causing defoliation of whole fields in a couple of weeks. Individual outbreaks can be controlled with fungicides, but the organism is here to stay because it can infect 100 other plant species besides soybeans. The annual U.S. soybean crop is worth $18 billion.

The rice blast fungus, *Magnaporthe grisea*, is a developing problem wherever rice is grown.[44] This disease was described in China in the seventeenth century, but has since moved around the planet and destroys enough rice to feed 60 million people.[45] A fungal overachiever, *Magnaporthe* infects many other crops including wheat, barley, millet, corn, and sugarcane. Plant breeders have a continuous output of new rice cultivars with enhanced resistance to blast, but these have a very limited shelf life because the fungus is unusually good at spawning new races that overcome the plant's defenses. Unlike the rusts, new races of this pathogen evolve without regular sexual reproduction through a process called parasexual recombination. This onanistic trick is also effective at allowing the fungus to thwart new fungicides. On an optimistic note, the recent sequencing of the entire genome of the rice blast fungus is a phenomenal scientific achievement, and other pathogens are being laid bare in similarly exquisite detail.[46] Researchers hope that the publication of these encyclopedias of data will lead to the development of new approaches to controlling epidemic diseases. Among the interesting results that might emerge from the genomic research is the identification of suites of genes that determine why a particular fungus kills plants. This will be investigated by comparing the genomes of fungi that cause disease with fungi that feed on plant debris in soils. But while a cluster of genes may equip an Asian fungus with the tools that it needs to penetrate a rice leaf, the introduction of its spores to an American rice field has nothing to do with fungal genetics. Genetic research can't solve problems rooted in global commerce.

Speaking technically, rice blast, like many of the fungi in this book, is a very dangerous microbe. In fact, it is so destructive that the United States government conducted trials on blast in the 1940s as a biological warfare agent.[47] Experiments with spores of blast, and another rice pathogen, *Cochliobolus miyabeanus*, which causes brown spot,[48] weren't very promising, but efforts to use fungi as weapons continued through the 1950s and 1960s. Interest in rice blast grew with American involvement in Korea, and elsewhere in Southeast Asia, and researchers viewed potato blight and stem rust of cereals as diseases that could be potentially valuable weapons. A variety of delivery systems were tested. The most bizarre device was a feather bomb—a huge canister filled with spores dusted on turkey feathers.[49] An airborne burst would scatter the infectious feathers, which would then descend slowly over a wide area to activate the epidemic. Imitating a secret weapons program operated by the Japanese during the Second World War, American researchers studied the effectiveness of hydrogen-filled balloons for transporting these canisters to their targets (fig. 8.4).[50] Stockpiles

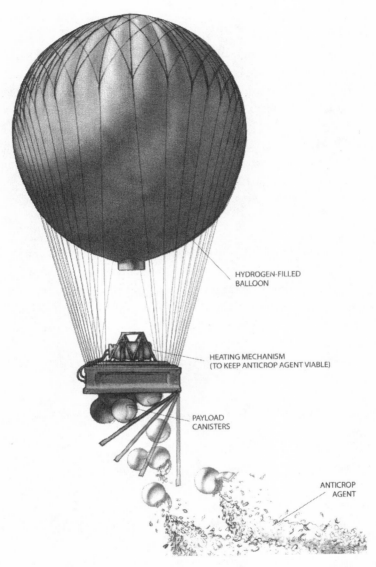

HYDROGEN-FILLED
BALLOON

HEATING MECHANISM
(TO KEEP ANTICROP AGENT VIABLE)

PAYLOAD
CANISTERS

ANTICROP
AGENT

Fig. 8.4   Design for hydrogen-filled biowarfare balloon developed by the U.S. military for the dissemination of anticrop biological agents. The "bombs" contain feathers bearing fungal spores. From P. Rogers, S. Whitby, and M. Dando, *Scientific American* 280, 70–75 (1999), with permission.

of rust and blast spores were replaced on a regular schedule to maintain viability. More than 30 tonnes of stem rust spores were produced, which was enough to infect every wheat plant on the planet. The Soviet Union was interested in the same diseases, but rather than stockpiling spores, they planned to produce masses of the infectious cells in a short period of time and deliver them in intercontinental ballistic missiles.

Despite the lobbying efforts of military planners who viewed epidemic crop diseases as the icing on the cake for the sickened survivors of nuclear holocaust, the American program was dismantled in 1969 at the behest of an executive order from President Nixon. The official justification for this unilateral action was that biological agents were of limited military value. The truth was trickier. Nixon recognized that continued development of biological agents by the United States might stimulate copycat nations to do the same. Few nations possessed the technology to build atomic bombs, but a group of committed peasants could grow bacteria or fungi in a tub. The Soviet military didn't follow this logic and continued their military investment in plant pathology until the 1980s.

No mention of biological warfare is complete without a brief mention of Saddam Hussein. His use of poison gas against the Kurdish population of northern Iraq is well documented, and there is evidence that fungal toxins were used in the same region. After the Gulf War in 1991, details of Iraq's chemical and biological warfare program were compiled by United Nations inspectors. Iraq disclosed that aflatoxins were produced at the Agricultural and Water Research Station, also known as the Fudaliyah Site, east of Baghdad in the 1990s.[51] Aflatoxins produced by species of *Aspergillus* are important in the spoilage of peanuts and other foods. They are potent carcinogens and cause liver cancer in laboratory animals, but their deployment as battlefield weapons seems illogical because these effects are unlikely to become apparent for several months or even years. According to Jeffrey Goldberg, a staff writer at *The New Yorker*, arms inspectors joked that "aflatoxin would stop a lieutenant from making colonel, but it would not stop soldiers from advancing across a battlefield."[52] Nevertheless, Iraqis claimed to have stockpiled 2,220 liters of the stuff and deployed this in the warheads of Al Hussein missiles and other munitions.[53] The Iraqis also recognized the value of crop pathogens and are known to have worked with *Tilletia tritici* and *Tilletia laevis*, the covered smuts. The likely target was Iran, but by their own admission they were unsuccessful in developing smut bombs. As we are all aware, there were no traces of biological weapons, mycological or otherwise, in the aftermath of the 2003 invasion of Iraq.

The United States government has actively pursued the use of pathogenic fungi in the control of coca, cannabis, and opium poppies. Fungi used in this fashion are called mycoherbicides. *Fusarium oxysporum* is active against all three narcotic plants, and another species, *Pleospora papaveracea*, is studied for the control of poppies. There are, understandably, serious political hurdles that complicate the use of American mycoherbicides in coca-growing countries in South America and in the poppy fields of Afghanistan. One fly in the ointment is the potential vulnerability of staple crops to the fungi used in drug control. Homeland security in the United States is another issue. The attacks on September 11, 2001 provoked a lot of interest in the threat to U.S. agriculture posed by terrorists. Plant pathologists are concerned about a number of pathogenic fungi, including cereal rusts and smuts, potato blight, and rice blast. The same fungi are an obvious threat to agriculture in other parts of the world, but each country has its specific home-grown targets. Earlier in the book we considered the "disease-free" cacao crops of West Africa and the rubber plantations of Asia. Fungi are ready to oblige with an epidemic among these delicate monocultures as soon as someone acts as the appropriate vector.

಑ఎ

At the end of a talk by SOD expert David Rizzo in 2002, I thanked him for the most depressing seminar that I had ever heard. *Phytophthora ramorum*, *Phytophthora cinnamomi*, and their relatives continue to extend their geographical range and to suck the life from an ever greater variety of plant species. The emergence of new hybrid species of *Phytophthora* with the kind of apocalyptic host ranges seen in the SOD and jarrah dieback pathogens is perhaps the most unsettling prospect for plant pathologists, as well as the rest of us. Indeed, the impact of these unstoppable microbes against a sufficiently broad range of plant species, or just a few key plant species, could signal the end of civilization. Let me be more specific. A microbe that behaved like *Phytophthora ramorum* but destroyed cereal plants would see most of us to our graves. It's possible that something similar happened to another dominant group of animals a very long time ago.

The Global Impact of Fungal Disease

Massive extinctions occur once in a while. Most of the planet's species were eliminated at the end of the Permian, and at least half those living at the end of the Cretaceous also packed their bags in a hurry. Asteroid strikes or volcanic eruptions are prime suspects as the triggering events in both cases. If an asteroid strike, or volcano, threw enough dust into the atmosphere to plunge the planet into semidarkness for a while, the effect on plant productivity would have been fast and furious. The resulting collapse of food chains on land and in the oceans would then explain the deaths of entire orders of animals. Geologists have discovered compelling evidence of plant extinction at the end of the Cretaceous, the so-called K-T boundary, in the form of a botanical and mycological anomaly in the fossil record. Coal deposits from New Zealand dating to just before the K-T event contain a typical mixture of spores and pollen grains from 80 or so plant species. Above this, plant spores and pollen grains are absent from a thin layer of sediment that is full of fungal spores and fragments of hyphae.[54] The fungal layer is only 4 millimeters deep, suggesting that it was deposited in no more than a few years, an instant in geological time. We can't be certain what happened, of course, but it seems likely that the masses of fungal spores reflect the putrefying mountain of plant life extinguished by the K-T calamity. Later, the rocks show an abundance of fern spores, suggesting that the postapocalyptic landscape was quickly regreened.[55]

All of this meshes nicely with the hypothesis that the K-T event destroyed the vegetables that fueled the global food chains, and that the dinosaurs (along with many other animals) starved to death. But there are other scenarios. My friend Arturo Casadevall has a provocative mycological solution to the extinction of the dinosaurs.[56] Of all of the species of fungi, very few possess specific adaptations that suit them for growing in animal tissues. To invade our tissues, a fungus must be able to grow at the moderately elevated temperature controlled by our onboard thermostats. This requirement keeps fungi adapted to cool conditions out of the medical textbooks. [57] Protection from microbial diseases, particularly those caused by fungi, might have served as a powerful stimulus for the evolution of endothermy (or warm-bloodedness). Arturo theorizes that the unprecedented accumulation of fungal spores after the global plant die-off would have altered the usual balance of power, delivering such massive concentrations of spores into the lungs of the dinosaurs that their immune defenses were overwhelmed.[58] If the body temperature of dinosaurs was lower than that of a mammal, then these reptiles might well have had greater susceptibility to fungal infection. Warm-blooded mammals would have had a physiological advantage in the deluge of fungal spores, their tissues being too warm for the germinating spores. Numerous criticisms can be launched against this hypothesis, not least of which is the ongoing dispute about the cold- or warm-blooded nature of the dinosaurs. Nevertheless, the idea that extinction of the dinosaurs might have been caused by fungal infection rather than by starvation is very provocative.

Conifers and seed ferns dominated the Cretaceous flora. After the K-T event, the conifers declined, the flowering plants diversified, and the seed ferns disappeared. The flowering plants are thought to have evolved from another order of seed plants called the Bennettitales, or cycadeoids, which arrive and depart from the fossil record with the dinosaurs.[59] Regardless of the means of their death, fungi would have decomposed the tissues of any plants that accumulated at the end of the Cretaceous, but perhaps the link between plant extinction and fungal proliferation could be more direct than Casadevall's scenario. Could a super-pathogen that took advantage of the unrelenting K-T dusk have driven the Cretaceous plants to extinction?[60] The defenses of the prehistoric plants would have been weakened by light limitation, and their stress would have been amplified by the continual deposition of dust (from an asteroid impact or a volcano) on their leaves. Nobody knows what the vegetarian dinosaurs ate, but the catholic host range of some contemporary Phytophthoras and rusts shows, at least in principle, that a single microbial agent could eliminate all of their food supply. Like today's animal inhabitants of the Australian forests, dinosaurs couldn't survive a plague that severed their food chains. If the elimination of the dinosaurs furnished the ecological opportunity for the diversification of mammals, then it might be said for the purpose of popularizing this fungal hypothesis that we owe our origins to the pathogens that decimated the Cretaceous forests. Perhaps pathogenic fungi aren't so bad after all.

While cataclysmic interactions between fungi and plants have presumably occurred throughout biological history, this doesn't justify a Panglossian view of plant pathology. Our assault upon the biosphere has worsened the impact of fungal epidemics and has greatly increased their frequency. Consider, once more, that many of the stories in this book—coffee rust, witches' broom of cacao, rubber blight, potato blight, and epidemic cereal diseases—are rooted in our insistence on growing monocultures of crop plants. In the growing number of diseases like SOD and jarrah dieback where pathogens attack hosts that seem blessed with genetic diversity, human intervention comes in the form of habitat disturbance and the global movement of plant material. With human population soaring beyond 6 billion, and current projections showing us at 9 billion by the middle of this century, continued reliance on monocultures and the accelerated erosion of natural habitats seem inescapable. When we consider the varied ways in which *Homo sapiens* could be extinguished, blights, rusts, and rots must be on the list. The fungi are everywhere and will outlive us by an eternity.

Notes

Chapter 1

1. H. W. Merkel, *Annual Report of the New York Zoological Society* 10, 97–103 (1905), pp. 102–103.

2. W. A. Murrill, *Autobiography* (W. A. Murrill, 1945), p. 70.

3. W. A. Murrill, *Journal of the New York Botanical Garden* 7, 143–153 (1906); W. A. Murrill, *Torreya* 6, 186–189 (1906).

4. W. A. Murrill, *Torreya* 8, 111–112 (1908).

5. W. A. Murrill, *Journal of the New York Botanical Garden* 9, 23–31 (1908).

6. Measurements of enormous American chestnuts in Massachusetts, attributed to Emerson, are given in material added to later editions of an English translation of the classical account of American forests by François André Michaux: F. A. Michaux, *The North American Sylva*, vol. 3 (Philadelphia: D. Rice and A. N. Hart, 1857). Michaux wrote that a European chestnut on Mount Etna in Italy was said to be 49 meters (160 feet) in circumference "and large enough to shelter one hundred men on horseback beneath its branches" (p. 12).

7. G. H. Hepting, *Journal of Forest History* 18, 60–67 (1974).

8. S. Anagnostakis, *Biological Invasions* 3, 245–254 (2001).

9. European chestnuts are still used in tanneries in Italy.

10. *The Conference Called by the Governor of Pennsylvania to Consider Ways and Means for Preventing the Spread of the Chestnut Tree Bark Disease*, reported by Guilbert and Lewis of Philadelphia, PA (Harrisburg, PA: C. E. Aughinbaugh, 1912).

11. The Blight Commission derived the figure of $70 million by combining the value of the timber and nut crop. To estimate the number of trees, they used 1909 figures for forest cover from Pennsylvania's Auditor General and assumed that one-fifth of the forest was occupied by chestnuts.

12. The disease spread in all directions across the range of the chestnut, eventually killing trees from Maine to Alabama.

13. *Conference Called by the Governor* (n. 10), p. 17.

14. *Conference Called by the Governor* (n. 10), pp. 116, 223.

15. *Conference Called by the Governor* (n. 10), p. 40.

16. *Conference Called by the Governor* (n. 10), pp. 20, 41, 108. Hepting (n. 7). At the time of the conference, only $20,143 of the appropriation had been spent.

17. *Report of the Pennsylvania Chestnut Tree Blight Commission, July 1 to December 31, 1912* (Harrisburg, PA: C. E. Aughinbaugh, 1913).

18. Hepting (n. 7).

19. *American Forestry* 19, 556–558 (1913).

20. J. S. Holmes, in *Chestnut and the Chestnut Blight in North Carolina*, North Carolina Geological and Economic Survey Economic Paper 56 (Raleigh, NC; 1925), p. 6.

21. Although the fungus penetrates the sapwood, it doesn't stop water flow by accumulating in the conductive vessels, but does so indirectly by inducing the tree to form plugs called tyloses; W. C. Bramble, *American Journal of Botany* 25, 61–65 (1923). In a sense, the tree is injured by its own response to infection, much like a severe allergic reaction.

22. The term "canker" was used for cancer until seventeenth century, but since then it has been applied more often to tree diseases, ulcerous mouth sores, and equine illnesses.

23. Merkel (n. 1), p. 97.

24. Here's how I estimated the speed of the epidemic's spread. The disease was identified first in 1904 in New York City. The blight spread north and west, but traveled the greatest distance in a southwestward track. By 1940, the blight had spread to northern Alabama and Mississippi, and by 1950 to southern Mississippi. The distance from New York City to Jackson, Mississippi, as the spore-carrying bird flies, is approximately 1800 kilometers (1100 miles). This distance divided by 46 years provides an estimate of 39 kilometers per year, which equals 3.2 kilometers per month or 107 meters per day.

25. F. D. Heald and R. A. Studhalter, *Journal of Agricultural Research* 2, 405–422 (1914).

26. One bird that wasn't available for shooting in 1914 was the passenger pigeon. The last wild bird, later named Buttons, had been shot by 14-year-old Press Clay Southworth in Sargents, Ohio in 1900. The name Buttons refers to the eyes fitted to the stuffed bird by a taxidermist. Buttons is on display in the Ohio Histor-

ical Society Museum in Columbus. The vast flocks of this native species would certainly have spread the blight had it not been extinguished before the epidemic.

27. Mating involving three parents was described by S. Anagnostakis, *Genetics* 100, 413–416 (1982). *Cryphonectria* is also capable of self-fertilization, meaning that even in the absence of a compatible mate, the fungus can produce the sexual spores called ascospores. Even though this is quite a rare occurrence in the laboratory, it happens often in the wild.

28. I. S. Cunningham, *Frank N. Meyer, Plant Hunter in Asia* (Ames: The Iowa State University Press, 1984).

29. D. Fairchild, *Science* 38, 297–299 (1913), pp. 297–298.

30. C. L. Shear and N. E. Stevens, *Science* 43, 173–176 (1916).

31. S. Anagnostakis, *Chestnuts and the Introduction of Chestnut Blight*, Fact Sheet PP008 (New Haven, CT: The Connecticut Agricultural Experiment Station, 1997).

32. Anagnostakis (n. 31), explains that saplings of the Chinese chestnut shipped from a seed nursery in Maryland may have assisted the later spread of the disease in Southern states.

33. Sprouts grow from woody bumps at the base of the tree called the root collar. Ellen Mason Exum described the chestnut as a "tree in a coma"; *American Forests*, November/December 1992, 20–25, 59–60.

34. C. Maynard et al., *Journal of the American Chestnut Foundation* XII, 2, 40–56 (1998).

35. Frogs produce corkscrew-shaped peptides, called magainins, in their skin and intestines that offer remarkable resistance to infection by fungi, bacteria, and viruses. Though far from likely, it is conceivable that a tree coerced to produce these broad-spectrum antimicrobial peptides might obtain some protection from chestnut blight. Besides frog genes, blight researchers are also looking at the protective effects of genes introduced from wheat and amaranth.

36. Disease-free European chestnuts still thrive in the virgin forest of Russia's Caucasus State Biosphere Preserve; F. Paillet, *The Journal of the American Chestnut Foundation* 9, 48–59 (1995–1996).

37. S. Anagnostakis, *Mycologia* 79, 23–37 (1987).

38. A. L. Dawe and D. L. Nuss, *Annual Review of Genetics* 35, 1–29 (2001).

39. M. G. Milgroom and P. Cortesi, *Annual Review of Phytopathology* 42, 311–338 (2004).

40. Although hypovirulence doesn't offer a lot of hope for restoring the American chestnut, European trees continue to show the disease resistance discovered

after the initial wave of damage. The reason for the different outcome isn't clear. Sandra Anagnostakis raised the possibility that Italian trees owe their resistance to a practice of grafting *Castanea sativa* onto a chestnut variety called "Marrone." Marrone was selected by Turkish monks in the twelfth century, and Anagnostakis has pondered whether the monks introduced any "Asian blood" into the variety by hybridizing European and Asian trees; *The Journal of the American Chestnut Foundation* 8, 10–11 (1994).

41. Established in 1983, the American Chestnut Foundation has more than 5,000 members. Honorary Board Directors include President Jimmy Carter and Nobel Peace Prize recipient Norman Borlaug; www.acf.org.

42. C. E. Little, *The Dying of the Trees. The Pandemic in America's Forests* (New York: Viking, 1995).

Chapter 2

1. *Drayton St Leonard: Our Village* (Drayton St Leonard, Oxfordshire: Drayton St Leonard Historical Society, 2000).

2. C. M. Brasier, *Plant Pathology* 39, 5–16 (1990).

3. The first records of the epidemic may be traced to 1912, but the intervention of the war may explain why these sightings did not become part of a seamless record of the epidemic's progress.

4. F. W. Holmes and H. M. Heybroek, *Dutch Elm Disease—The Early Papers. Selected Works of Seven Dutch Women Phytopathologists* (St. Paul, MN: The American Phytopathological Society, 1990).

5. The description is translated from Schwartz's doctoral dissertation published in 1922 and is taken from Holmes and Heybroek (n. 4), p. 53.

6. Buisman named the sexual stage *Ceratostomella ulmi*. According to the rules of taxonomy this became the preferred name for the pathogen, replacing *Graphium ulmi*. C. J. Buisman, *Tijdschrift over Plantenziekten* 38, 1–5, plates I–III (1932).

7. Charles F. Irish was the arborist who sent the twigs to Buisman. His company illustrated the disease on a poster published in 1930, and Curtis May who examined infected trees in Cleveland and Cincinnati reported his findings in *Science* 72, 142–143 (1930).

8. Richard J. Campana illuminates the history of research on Dutch elm disease in his book, *Arboriculture: History and Development in North America* (East Lansing, MI: Michigan State University Press, 1999).

9. B. Clouston and K. Stansfield, *After the Elm* (London: William Heinemann, 1979).

10. J. A. Byers, P. Svhira, and C. S. Koehler, *Journal of Arboriculture* 6, 245–246 (1980). The descriptions of the odor of these chemicals is taken from http://www.thegoodscentscompany.com/rawmatex.html.

11. T. J. Cobbe, unpublished manuscript, courtesy of the Willard Sherman Turrell Herbarium, Miami University, Oxford, OH.

12. A single diseased elm was found in Cleveland in 1933, two more in 1934, and others in 1935. All were cut and burned.

13. Campana (n. 8).

14. I was amused by a misspelling in a 1930 article about George Keely in the *Hamilton Evening News* (November 3): "Behind the foliage of Oxford town always, village historians point out, lurks the spirit of their sewer, Dr. George W. Keely, who walked the wide streets long before the Civil war, and planned for leafy beauty for generations to come."

15. R. Wolkomir and C. Davidson, *Smithsonian* 29(3) (1998), p. 43

16. In America, sycamore is the common name for *Platanus occidentalis*, which is known as the plane tree in Europe.

17. W. C. Bramble, *American Journal of Botany* 25, 61–65 (1923).

18. C. G. Bowden et al., *Molecular Plant-Microbe Interactions* 9, 556–564 (1996). Subsequent work suggests that a coating of cerato-ulmin on the budding yeast cells increases their resistance to dessication. This may be critical when the fungal cells are transported on the bark beetles.

19. A. Solla and L. Gil, *Forest Pathology* 32, 123–134 (2002). Vessels are composed of short pipes arranged end to end and separated by membranes. These membranes, and the rest of the vessel walls, are traversed by pores. The size of these pores has a major influence on the rate of water movement through the xylem and the vulnerability of the vessels to cavitation. Tyloses block these pores.

20. Clouston and Stansfield (n. 9), p. 66.

21. G. Wilkinson, *Epitaph for the Elm* (London: Hutchinson, 1978), p. 66. The death toll in Britain after 30 years of the Dutch elm disease epidemic is estimated at 25 million trees.

22. C. M. Brasier and J. N. Gibbs, *Nature* 242, 607–609 (1973).

23. All the pertinent literature is cited in Braiser (n. 2).

24. Brasier believes that the Eurasian aggressive race appeared first in Romania in the 1940s. The pair of aggressive races are considered to be a subspecies of *Ophiostoma ulmi*, called *Ophiostoma novo-ulmi*.

25. Edwin J. Butler (1874–1943) served as Imperial Mycologist in India and later became the Director of the Imperial Mycological Institute at Kew Botanic Gardens. Butler outlined his wicker basket theory in a letter written in 1934 that was published in an article by G. P. Clinton and F. A. McCormick, *Connecticut Agricultural Experiment Station Bulletin* 389, 701–752 (1936).

26. J. G. Horsfall and E. B. Cowling, in *Plant Disease: An Advanced Treatise*, vol. 2, edited by J. G. Horsfall and E. B. Cowling (New York: Academic Press, 1978), p. 22.

27. J. N. Gibbs, *Phytopathology* 70 (1980), p. 699.

28. W. A. Watts, *Proceedings of the Linnean Society of London* 172, 33–38 (1961).

29. Dendrophilus, *Philosophical Magazine* 62, 252–254 (1823).

30. R. Mabey, *Flora Brittanica. The Concise Edition* (London: Chatto and Windus, 1998).

31. D. W. French, *History of Dutch Elm Disease in Minnesota*, Minnesota Report 229 (St. Paul,: Minnesota Agricultural Experiment Station, University of Minnesota, 1993).

32. Vaccination of elms with proteins that stimulate the tree defenses is a promising venture. When trees are injected with a specific protein that is normally generated by the fungus, they respond by releasing compounds called mansonones into their xylem vessels. Mansonones kill the fungus by shutting down its mitochondria. The treatment, developed by Martin Hubbes at the University of Toronto, is marketed as ELMguard. A. Coghlan, *New Scientist* 160, 7 (1998).

33. Estimates show that 70 percent of mature elms in the United States have been destroyed by Dutch elm disease. If each of the remaining seven million trees is valued at $3,000, the total stand is worth $21 billion, representing a tremendous incentive for researchers. Estimates based on rounding of numbers offered by D. Sawyer, *The Forestry Chronicle* 77, 961 (2001).

34. A. DePalma, *New York Times* (May 7, 2004), pp. A1, C16.

35. L. Gil et al., *Nature* 431, 1053 (2004).

36. *Lucius Junius Moderatus Columella: On Agriculture X–XII, On Trees*, trans. E. S. Forster and E. Heffner (Cambridge, MA: Harvard University Press, 1955). Reference to Atinian elms appears in Columella's discourse *On Trees* XV, XVI. Pliny the Elder wasn't enthralled by the Atinian variety, saying that it bore too many leaves (and would, presumably, shade the vines). But in a compromise with Columella's advice, he suggested that Atinian elms should be interspersed

with other varieties of elm in the arbustum; *Pliny, Natural History*, vol. V, trans. H. Rackham (Cambridge, MA: Harvard University Press, 1950), Book XVII, XXXV, 200.

37. Because the double-stranded RNA "d-factors" that induce hypovirulence in *Ophiostoma* lack the protein coat that equips most viruses, these antifungal entities are often referred to as "viruslike agents." S. Pain, *New Scientist* 153, 26–30 (1997).

38. M. Arnold, Rugby Chapel, November 1857 (1867).

Chapter 3

1. M. Pendergrast, *Uncommon Grounds. The History of Coffee and How It Transformed Our World* (New York: Basic Books, 1999).

2. D. Lorenzetti and L. R. Lorenzetti, *The Birth of Coffee* (New York: Clarkson Potter, 2000).

3. The figure of 6–7 billion kilograms (6 million tonnes), from the International Coffee Organization (www.ico.org), includes robusta beans from the West African native, *Coffea canephora* (robusta coffee). To comprehend this quantity, it may be helpful to consider that the largest ocean liner, the *Queen Mary 2*, weighs 150,000 tonnes—the weight of the annual coffee crop is, therefore, equivalent to 40 or more vessels of this size.

4. www.lankalibrary.com.

5. Ceylon's nineteenth-century coffee crop matches today's annual production in Venezuela and represents approximately 1 percent of global production (www.ico.org).

6. Against a background of continuing deforestation, 830 of the 3,000 species of flowering plants in Sri Lanka are endemic, and 93 percent of these are found in the remaining rainforests that account for 2 percent of the land area. N. M. Collins, J. A. Sayer, and T. C. Whitmore, eds., *The Conservation Atlas of Tropical Forests of Asia and the Pacific* (New York: Simon and Shuster, 1991).

7. W. Knighton, *Forest Life in Ceylon*, 2 vols. (London: Hurst and Blackett, 1854), vol. I, 120–121, 283–285.

8. E. L. Arnold, *Coffee: Its Cultivation and Profit* (London: W. B. Whittingham & Co., 1886), p. 48. To balance the negative characterization of his workers, Arnold flexed his egalitarian muscles in this book by writing: "The coolie's interest is the planter's. He should be lodged well, fed sufficiently whatever the price of grain is, and kindly treated" (p. 46).

9. The end of slavery in Britain's colonies was mandated by the Abolition of Slavery Act of 1833, but until section 64 of the act was repealed, Ceylon, St. Helena, and British India were excepted from the ban.

10. S. Baker, *The Rifle and the Hound in Ceylon. Stories from the Field, 1845 to 1853* (London: Longmans, 1854), is not for girly men like myself. Referring to elephant hunting, Baker wrote: "I have shot them successfully both in the brain and in the shoulder, and where the character of the country admits an approach to within ten paces, I prefer the Ceylon method of aiming either at the temple or behind the ear" (p. vii). Although hunting wasn't the chief concern of his next book, *Eight Years' Wanderings in Ceylon* (1855), the title page was accompanied by an etching of a hunter and his dogs chasing a terrified stag off a cliff.

11. Despite his views of the difficulties of farming in Ceylon, Baker's optimism about coffee-growing is quite remarkable. The second edition of *Eight Years* was published in 1884, in the midst of the extinction of coffee in Ceylon, and yet Baker wrote that decades of development had all but extinguished the forests of the hill country, and that every acre would soon be producing the crop. Baker's apparent ignorance of the problem is made more surprising by his interest in mycology during his retirement.

12. George Henry Kendrick Thwaites (1811–1882) was an accountant with a passion for botany who discovered that diatoms were photosynthetic algae, rather than animals (as had been believed). He became a lecturer in Bristol before 30 years service as superintendent, then director, of the Botanical Gardens in Peradeniya. He was one of several early correspondents with Darwin following the publication of *On the Origin of Species*. Thwaites disputed the evolutionary relevance of the elaborate patterning of diatom shells; F. Darwin (Editor), *More Letters of Charles Darwin*, vol, 1, letter 97, March 21, 1860 (London: John Murray, 1903).

13. There is disagreement in the literature about the earliest report of coffee rust. Lester Arnold (n. 8) wrote that the disease had been known in Ceylon before 1840, but he was probably referring to the ubiquitous weblike growth of other fungi on coffee leaves in the humid climate. Some sources refer to the disease in Ceylon in 1861, and others state that an unnamed British explorer noted the disease in the Lake Victoria region of Africa in the same year. Neither appears to be true. While searching for the source of the Nile in 1862, John Speke and James Grant described wild coffee trees growing in eastern Kenya and described the practice of native people who sun-dried unripe berries and ate them as stimulants. The species was identified as *Coffea arabica*, but these plants

were probably the rust-resistant *Coffea canephora*, because *arabica* isn't a native of Kenya. This was the conclusion reached by F. L. Wellman in *Coffee: Botany, Cultivation, and Utilization* (London: Leonard Hill; New York: Interscience, 1961). But despite his careful sieving of the literature, Wellman seems to have corrupted the story of the rust's appearance by infecting the coffee plants seen by the Speke expedition and altering the year to 1861 (e.g., F. L. Wellman, *Foreign Agriculture* 17, 47–52 [1952]). Wellman is the primary source that spawned the oft-quoted discovery of the disease in Africa, and others amplified the error by transferring the fallacious African sighting to the plantations of Ceylon. More interesting, Speke and Grant made an unquestionably important discovery in 1862 when they proved that the Nile issues from Lake Victoria at Ripon Falls in Uganda. In 1863, they crossed paths with Baker and von Sass in southern Sudan and gave them information that helped the Baker expedition find Albert Nyanza, or Lake Albert.

14. M. J. Berkeley, *The Gardeners' Chronicle and Agricultural Gazette* November 6, 1869, p. 1157.

15. J. Nietner, *The Coffee Tree and Its Enemies: Being Observations on the Natural History of the Enemies of the Coffee Tree in Ceylon*, 2nd edition (Colombo, Ceylon: A. M. & J. Ferguson, Ceylon Observer Press, 1880).

16. Arnold (n. 8), p. 119.

17. Nietner (n. 15), p. 20.

18. E. C. Large, *The Advance of the Fungi* (New York: Henry Holt & Company, 1940), p. 198.

19. P. Ayres, *Harry Marshall Ward and the Fungal Thread of Death* (St. Paul, MN: The American Phytopathological Society, 2005). This is a masterly and detailed volume of scientific biography. For a briefer analysis of Ward's career, refer to a beautifully written piece by G. C. Ainsworth in *Annual Review of Phytopathology* 32, 21–25 (1994).

20. H. M. Ward, *Journal of the Linnean Society. Botany* XIX, 299–335 (1882).

21. R. W. Rayner, *Nature* 191, 725 (1961).

22. A young coffee tree produces 200–400 leaves.

23. Large (n. 18), p. 203.

24. H. M. Ward, *Quarterly Journal of Microscopical Science* XXII, 1–11 (1882). Berkeley was first to illustrate the teliospores in his 1869 description of the fungus (fig. 3.2), but did not recognize that they were a distinct spore type.

25. N. P. Money, *Mr. Bloomfield's Orchard. The Mysterious World of Mushrooms, Molds, and Mycologists* (New York: Oxford University Press, 2002).

26. Daniel Morris, Ward's predecessor, believed that coffee rust worked in the same way as hop mildew, forming a web of hyphal threads over the leaf surface.

27. The extract from a letter to the *Ceylon Observer* (December 15, 1880) is quoted by F. B. Thurber, *Coffee: From Plantation to Cup. A Brief History of Coffee Production and Consumption* (New York: American Grocer Publishing Association, 1881), p. 95.

28. J. Eriksson, *Comptes Rendus* 124, 475–477 (1897). Eriksson didn't offer illustrations of his *corpuscules spéciaux* in this note, saving disclosure for a separate paper in 1902.

29. The same idea prevails when we think of most of the infectious diseases of humans. Viruses do offer a bit of an exception to this because some emerge from our cells, having hidden as part of our genome, and of our ancestors' genomes, for a very long time. But this is putting oncogenes into the picture, and I don't need to complicate things to explain why Jakob Eriksson's idea was silly.

30. Eriksson didn't have any time for spontaneous generation (to which the quote from Nicander refers), but his ideas were disturbingly Medieval in their reliance upon faith rather than empirical evidence. For an introduction to the development of classical ideas about the nature of fungi, an old paper by A. H. R. Buller remains an entertaining and informative read: *Transactions of the Royal Society of Canada* (Series III) 9, 1–25, & plates I–IV (1915).

31. Eriksson was a stupendously obstinate scientist who stuck to his mycoplasm theory for decades after Ward had clearly shown it was false. The concept appeared in Eriksson 's influential textbook, *Fungous Diseases of Plants* (London: Baillière, Tindall and Cox, 1930), on the still young science of plant pathology. But he betrayed his uneasiness by omitting mention of coffee rust and Ward in the lengthy tome and by relegating De Bary to a brief, unfavorable comment and single footnoted reference. Eriksson's failure to highlight the brilliance of De Bary, and his omission of Ward's work on coffee rust, is comparable to penning a textbook on American history and neglecting George Washington and Thomas Jefferson. Despite his lengthy lapse in judgment, Eriksson is considered one of the great figures in plant pathology for his discovery of genetically distinct races within a single species of rust.

32. H. M. Ward, *Philosophical Transactions of the Royal Society, London, B* 196, 29–46 (1903).

33. F. L. Wellman, *Plant Disease Reporter* 54, 539–541 (1970), p. 539.

34. W. H. Ukers, *All About Tea*, 2 vols. (New York: The Tea and Coffee Trade Journal Company, 1935), vol. I, 177.

35. www.teamuse.com.

36. Pendergrast (n. 1).

37. E. Schieber, *Annual Review of Phytopathology* 10, 491–510 (1972).

38. J. K. M. Brown and M. S. Hovmøller, *Science* 297, 537–541 (2002); E. A. Shinn, D. W. Griffin, and D. B. Seba, *Archives of Environmental Health* 58, 498–504 (2003).

39. E. Schieber (n. 37), p. 493.

40. J. Bowden et al., *Nature* 229, 500–501 (1971).

41. R. P. Scheffer, *The Nature of Disease in Plants* (Cambridge: Cambridge University Press, 1997).

42. F. Anthony et al., *Theoretical and Applied Genetics* 104, 894–900 (2002).

43. F. L. Wellman (n. 13). Wellman offered a painstaking exploration of the origins of cultivated coffee in his 1961 book. He acknowledged that parts of the story he developed are unclear. It would be interesting to know, for example, whether the British developed their Ceylonese monoculture from plants introduced by the Portugese or the Dutch, or brought their own plants originating from the tree grown in Paris.

44. Robusta coffee is indigenous to the equatorial rainforests of Africa and accounts for less than one-fourth of the global coffee production (www.coffeeresearch.com). Some countries specialize in robusta cultivation; for example, it accounts for 91 percent of Ugandan coffee, and Vietnam produces nothing else.

45. Rotational programs in which coffee trees are replaced every seven or eight years have been effective at curtailing losses due to rust in some African countries and have reduced the quantity of chemicals sprayed on the plants.

46. The breeding of coffee varieties that thrive under the uninterrupted glare of tropical sunlight has led to the expansion of sun growing since the 1970s.

47. E. Fawole, *Nigerian Journal Of Tree Crop Research* 3, 64–70 (1999).

48. The estimated consumption of 3,300 cups per second cited on various web sites is difficult to verify; 3,300 cups per second corresponds to 100 billion per year. At annual consumer spending of $70 billion (www.ico.org), this means that the average cup costs 70 cents. Since home brewers in the United States, and coffee drinkers in poorer countries, spend considerably less on a cup, the 70-cent average must be heavily biased by the cost of those frothy concoctions at Starbucks and sales of the paraphernalia associated with coffee brewing and drinking.

Chapter 4

1. www.scharffenberger.com.

2. J. H. Hart, *Cacao: A Treatise on the Cultivation and Curing of Cacao*, 2nd ed. (Port-of-Spain, Trinidad: Mirror Office, 1900), p. 61.

3. J. H. Hart, *Cacao: A Manual on the Cultivation and Curing of Cacao* (London: Duckworth & Co., 1911).

4. Witches' broom was reported in the same year in Surinam.

5. Hart (n. 3), p. 76; quotes from L. A. A. De Verteuil, *Trinidad: Its Geography, Natural Resources, Administration, Present Condition, and Prospects*, 2nd ed. (London: Cassell & Co., 1884).

6. Some scholars have argued that cacao originated in Central America and was introduced to the Amazon basin by aboriginal cultures that migrated southward; P. Sanchez and K. Jaffe, *Interciencia* 17, 28–34 (1992).

7. A. M. Young, *The Chocolate Tree. A Natural History of Cacao* (Washington, D.C.: Smithsonian Institution Press, 1994), p. 83.

8. Although limited pollination reduces the number of flowers that bear fruit, there is a limit to the number of pods that a single tree can support: 30–40 is the average. For this reason, a natural thinning process called "cherelle wilt" causes many immature pods to shrivel up and die. Cherelle wilt would, therefore, limit fruit production even if pollination could be stimulated. Young (n. 7).

9. R. A. Rice and R. Greenburg, *Ambio* 29, 167–173 (2000); see also the nontechnical essay by Rice and Greenburg in *Natural History*, July/August 2003, 36–43.

10. Cacao plantations have gobbled up more than 10 percent of the original forest in Côte d'Ivoire.

11. Statistics from the International Cocoa Organization web site: www.icco.org.

12. Today's top six cacao producers are, in descending order: Côte d'Ivoire, Ghana, Indonesia, Nigeria, Brazil, and Cameroon. These countries account for more than 80 percent of the world's total cacao production.

13. Young (n. 7).

14. R. E. D. Baker and P. Holliday, *Witches' broom disease of cacao (Marasmius perniciosus Stahel)*, Phytopathological Paper No. 2 (Kew, UK: The Commonwealth Mycological Institute, 1957).

15. J. Orchard et al., *Plant Pathology* 43, 65–72 (1994).

16. The coffee rust fungus, *Hemiliea*, produces the same type of spore, but they have no known function in the coffee disease (see chapter 3). The shared production of basidiospores between the cacao- and coffee-infecting fungi reflects a common evolutionary heritage that converges on an ancestor that lived more than 100 million years ago: both species belong to the taxonomic phylum called the Basidiomycota.

17. Gerold Stahel gave the first scientific description of the fungus, which he named *Marasmius perniciosus*; *Bulletin Department van den Landouw in Suriname* 33, 1–26 (1915). The nuances of the name change are explained by D. N. Pegler, *Kew Bulletin* 32, 731–736 (1977).

18. A related species, *Marasmius rotula*, looks similar to the horsehair parachute and goes by the common names pinwheel mushroom, collared parachute, collared horsehair, and little wheel toadstool. Yet another species, *Marasmius equicrinus*, is recognized as a problem in cacao nurseries. The fungus infects the young plants, causing a disease known as horsehair blight; Hart (n. 3). The same species also affects tea and nutmeg plants. In this case, the horsehair appearance is due to the development of skeins of black rhizomorphs (hairlike cables composed of bundles of hyphae) on leaf surfaces.

19. G. W. Griffiths et al., *New Zealand Journal of Botany* 41, 423–435 (2003). There is evidence that the *Crinipellis* found on lianas is genetically distinct from the fungus that attacks cacao. For this reason, plant pathologists refer to the L-biotype of *Crinipellis* on lianas, and the C-biotype that occurs on cacao. Additional biotypes colonize other plants.

20. The seeds of this tree are used to prepare a drink by the Talamanca Indians of Costa Rica.

21. Griffiths et al. (n. 19). In addition to proteins, cacao produces organic compounds called phytoalexins that protect plants against fungi. Phytoalexins are sometimes described as plant antibiotics.

22. P. Silva, *Cacau e lagartão ou vassoura-de-bruxa: registros efetuados por Alexandre Rodrigues Ferreira nos anos de 1785 a 1787 na Amazônia*, Boletim Técnico 146 (Bahia, Brazil: Centro de Pesquisas do Caccau, Bahia, Brazil, 1987).

23. C. J. J. van Hall, *Cocoa* (London: MacMillan and Co., 1914).

24. G. W. Padwick, *Losses caused by plant diseases in the colonies*, Phytopathological Papers No. 1 (Kew, UK: The Commonwealth Mycological Institute, 1956).

25. N. Asheshov, www.gci275.com/lives/country02.shtml.

26. H. C. Evans, in *Tropical Mycology*, vol. 2, *Micromycetes*, edited by R. Watling et al. (UK: CAB International, 2002), 83–112.

27. H. A. Laker and J. W. de Silva e Mota, *Cocoa Growers' Bulletin* 43, 45–57 (1990).

28. The distance between Manaus, in Amazonas, and the site where witches' broom eventually appeared in Bahia is 2,600 kilometers or 1,600 miles.

29. R. E. D. Baker and S. H. Crowdy, *Memoirs of the Imperial College of Tropical Agriculture* 7, 1–28 (1943); quote from Baker and Holliday (n. 14), p. 21.

30. H. C. Evans and R. W. Barreto, *Mycologist* 10, 58–61 (1996).

31. H. C. Evans, *Cocoa Growers' Bulletin* 32, 5–19 (1981).

32. J. L. Pereira et al., *Turrialba* 39, 459–461 (1989).

33. J. L. Pereira, L. C. C. de Almeida, and S. M. Santos, *Crop Protection* 15, 743–752 (1996).

34. T. Andebrhan et al., *European Journal of Plant Pathology* 105, 167–175 (1999).

35. P. Gadsby, *Discover*, August 2002, 64–71.

36. A. Bellos and G. Neale, *The Sunday Telegraph* (May 10, 1998), p. 25.

37. Sir William had arranged for the introduction of cacao pods from the island of São Tomé, but, unfortunately for his son's claim, they were not grown there until 1886.

38. M. Rosenblum, *Chocolate: A Bittersweet Saga of Dark and Light* (New York: North Point Press, 2005).

39. According to ICCO statistics (see n. 11), Côte d'Ivoire produced 1.3 million tonnes of cacao in 2002–2003, and Ghana produced 475,000 tonnes.

40. Production in Côte d'Ivoire diminished in 2004 due to conflicts between the government and rebel groups in the cacao-growing region, with exports falling below 1 million tonnes. An unforeseen response by Ghana, which processed a bumper crop approaching 600,000 tonnes, led to a crash in prices. The story is told by Rosenblum (n. 38).

41. Rosenblum (n. 38).

42. Hedger quote in Asheshov (n. 25).

43. C. A. Thorold, *Diseases of Cocoa* (Oxford: Oxford University Press, 1975).

44. According to 2001 statistics compiled by the American Phytopathological Society (www.apsnet.org), black pod, caused by *Phytophthora* species, reduced cacao production in Africa, Brazil, Asia, by 450,000 tonnes at a cost of

$423 million (based on estimated crop value of $940 per tonne); witches' broom decreased Latin American production by 250,000 tonnes, and cost $235 million; frosty pod rot, caused by *Crinipellis roreri*, reduced Latin American production by 30,000 tonnes and cost $47 million; vascular streak dieback, caused by *Oncobasidium theobromae*, reduced Asian production by 30,000 tonnes and cost $28 million, and swollen-shoot disease caused by a virus (referred to as a badnovirus) accounted for a loss of 50,000 tonnes of cacao in Africa at a cost of $28 million.

45. N. P. Money, *Mr. Bloomfield's Orchard: The Mysterious World of Mushrooms, Molds, and Mycologists* (New York: Oxford University Press, 2002).

46. The black pod pathogen can also survive on discarded pod husks and attack plants from these infectious bases.

47. M. D. Coffey, in *Phytophthora*, edited by J. A. Lucas et al. (Cambridge, UK: Cambridge University Press, 1991), 411–432.

48. S. A. Rudgard, A. C. Maddison, and T. Andebrhan, eds., *Disease Management in Cocoa. Comparative Epidemiology of Witches' Broom* (London: Chapman and Hall, 1993).

49. L. H. Purdy and R. A. Schmidt, *Annual Review of Phytopathology* 34, 573–594 (1996).

50. U. Krauss and W. Soberanis, *Biological Control* 22, 149–158 (2001).

51. S. Sanogo et al., *Phytopathology* 92, 1032–1037 (2002).

52. A. E. Arnold et al., *Proceedings of the National Academy of Sciences USA* 100, 15649–15654 (2003); see also an excellent overview of this area of research by K. Clay, *Nature* 427, 401–402 (2004).

53. H. C. Evans et al., *Mycologist* 16, 148–152.

54. To prove the validity of this statement, we need to know whether *Crinipellis roreri* engages in sexual reproduction (mating between compatible strains) before it generates the cells interpreted as basidia. This issue is discussed in H. C. Evans, K. Holmes, and A. P. Reid, *Plant Pathology* 52, 476–485 (2003).

55. H. C. Evans et al., *Cocoa Growers' Bulletin* 51, 7–22 (1998).

Chapter 5

1. Data for 2005 from www.rubberstudy.com, the web site of an intergovernmental body called the International Rubber Study Group based in London.

2. H. Evans, G. Buckland, and D. Lefer, *They Made America: Two Centuries of Innovators from the Steam Engine to the Search Engine* (New York: Little, Brown, 2004).

3. W. Dean, *Brazil and the Struggle for Rubber. A Study in Environmental History* (Cambridge, UK: Cambridge University Press, 1987).

4. V. Thomas et al., *Annals of Botany* 75, 421–426 (1995).

5. Mad Ridley's method involved slicing the bark in a series of overlapping Y-shaped cuts, to produce a striking herringbone pattern. If you're considering a new career as a rubber tapper, I recommend the web site of the International Rubber Research and Development Board (www.irrdb.org), where you'll find descriptions of alternative and experimental tapping methods.

6. R. M. Klein, *The Green World. An Introduction to Plants and People* (New York: Harper & Row, 1987).

7. H. A. Wickham, *Rough Notes of a Journey through the Wilderness: From Trinidad to Para, Brazil, By Way of the Great Cataracts of the Orinoco, Atabapo, and Rio Negro* (London: W. H. J. Carter, 1872).

8. H. A. Wickham, *On the Plantation, Cultivation, and Curing of Parà Indian rubber (Hevea brasiliensis): With an Account of Its Introduction from the West to the Eastern Tropics* (London: Kegan Paul, Trench, Trübner, 1908).

9. Wickham (n. 8), refers to "7,000 odd [plants] originally brought out of the forest by me" (p. 4). Dean (n. 3), cites mutually supportive references to 60,000 to 74,000 seeds from a variety of contemporary sources. Dean believes that Wickham reduced the number in his memoirs in light of the fact that fewer than 3,000 of his seeds germinated. The payment of £740 obviously accords with the fee of £10 per thousand.

10. J. Loadman, *Tears of the Tree. The Story of Rubber—A Modern Marvel* (Oxford: Oxford University Press, 2005). Warren Dean's book (n. 3) is another excellent resource for the Wickham controversy.

11. Wickham (n. 8), p. 54.

12. Dean (n. 3), p. 18.

13. Dean (n. 3), p. 21.

14. One of the problems with this fantasy is that had I lived in Victorian England I would undoubtedly have been serving Wickham his brandy or, worse still, cleaning the club bathrooms.

15. N. B. Ward, *On the Growth of Plants in Closely Glazed Cases*, 2nd ed. (London: John Van Voorst, 1852). Ward described the effect of warmth and

24. F. Fontana, *Observations on the Rust of Grain*, trans. P. P. Pirone, *Phytopathological Classics* 2 (1932).

25. J. Banks, *A Short Account of the Cause of the Disease in Corn, Called by the Framers the Blight, the Mildew, and the Rust* (London: W. Bulmer & Co., 1805), p. 10. Others guessed at the relationship between barberry and the wheat disease. In a letter to Banks in 1906 (*The Pamphleteer* 6, 415–419 [1815]), the British horticulturalist Thomas Knight wrote, "I was . . . much disposed to believe the parasitical plants of the same species, and that the difference in the form and size of the seed vessels [spore-producing structures of *Puccina*] arose only from the difference in nutriment they derived from the wheat, and from the acrid juice of the barberry" (p. 496). Knight also conducted inconclusive experimental infections of wheat plants. In later years he promoted the concept of disease-resistant varieties of cereals and reported that a fungal disease of fruit trees, called peach leaf curl, could be controlled by sprinkling trees with a mixture of lime and sulfur. In 1816, a Danish schoolteacher, Niels Scholer, infected rye plants with rust aeciospores from barberry. Ainsworth (n. 12) provides further details.

26. D. Isley, *One Hundred and One Botanists* (Ames: Iowa State University Press, 1994).

27. Harry Marshall-Ward, of coffee rust fame, wrote a superbly well-balanced and objective obituary of De Bary for *Nature* 37, 297–299 (1888), beginning with the unpleasant fact that Anton had died following a painful illness that led to "an operation which entailed the removal of parts of the face." For a more recent tribute to De Bary, see J. G. Horsfall and S. Wilhelm, *Annual Review of Phytopathology* 20, 27–32 (1982).

28. A. de Bary, *Untersuchungen uber die Brandpilze und die durch sie verursachten Krankheiten der Pflanzen mit Rucksicht auf das Getreide und andere Nutzpflanzen* (Berlin: G. W. F. Muller, 1853).

29. Tulasne and Tulasne (n. 14); Tulasne (n. 14).

30. L.-R. Tulasne and C. Tulasne, *Selecta Fungorum Carpologia*, 3 vols., trans. W. B. Grove, edited by A. H. R. Buller & C. L. Shear (Oxford: Clarendon Press, 1931).

31. Large (n. 10), p. 131.

32. Large (n. 10) details De Bary's advances in understanding the life cycle of *Puccinia graminis* and the gaps that remained. The solution to the function of the spermatia by John Craigie and A. H. R. Buller in the 1920s is discussed in N. P. Money, *Mr. Bloomfield's Orchard: The Mysterious World of Mushrooms*,

Molds, and Mycologists (New York: Oxford University Press, 2002), 103–104, 176–180.

33. Large (n. 10).

34. A. de Bary, *Comparative Morphology of the Fungi Mycetozoa and Bacteria*, trans. H. E. F. Garnsey, revised by I. B. Balfour (Oxford: Clarendon Press, 1887).

35. De Bary's symbiosis embraced mutualistic relationships, such as the fungal-algal partnership in lichens ("mutualism" had been used prior to De Bary), commensalism (where one organism profits at no obvious cost to its partner), and parasitism (in which one party benefits to the detriment of the other).

36. The terms heteroecious (for more than one host) and autoecious (for a single host, which is exemplified by coffee rust) were invented by De Bary. I think them too complicated, and Ernest Large (n. 10) agrees with me: "They were forged and synthetic terms which went with fair English or plain German about as well as lumps of concrete in a meadow; they made a simple idea sound difficult, but in flinging them into the scientific literature, De Bary gave the verbose something to [pontificate] about" (p. 135).

37. M. Lutz et al., *Mycologia* 96, 614–626 (2004); R. Bauer, M. Lutz, and F. Oberwinkler, *Mycologia* 96, 960–967 (2004).

38. A. P. Roelfs, *Plant Disease* 66, 177–181 (1982).

39. E. C. Stakman, F. E. Kempton, and L. D. Hutton, *The Common Barberry and Black Stem Rust*, USDA Farmers' Bulletin 1544 (Washington, DC: U. S. Government Printing Office, 1927), pp. 3, 28.

40. A. P. Roelfs, *Canadian Journal of Plant Pathology* 11, 86–90 (1989).

41. Fungicides are, broadly speaking, a necessary evil. The danger that they pose to humans, and other animals, varies according to the chemical makeup of each fungicide and the amount that makes its way into the groundwater. Some fungicides stick to soil particles when they are washed from the surface of plants and do not contribute to the pollution of groundwater. Others are rapidly leached from soil. The rate of biodegradation of fungicides is determined by multiple factors.

Chapter 7

1. M. J. Berkeley, *Journal of the Horticultural Society of London* 1, 9–34, plates I–IV (1846), p. 9. Reprinted as M. J. Berkeley, *Observations, Botanical and*

Physiological, on the Potato Murrain (East Lansing, MI: American Phytopathological Society, 1948).

2. S. Buczacki, *Mycological Research* 105, 1283–1294 (2001).

3. Others reached the same conclusion about the fungal nature of the potato blight in the 1840s. A Belgian clergyman, Abbé Edouard van den Hecke, suggested that a fungus caused the disease and also referred to the mechanism of spore release. (Berkeley failed to describe zoospore release in his seminal paper, and De Bary has been credited with its discovery in the 1860s.) Charles Morren of Liège was an influential supporter of the fungal hypothesis. Bostonian James Teschemacher reached similar conclusions in 1844 during the earlier disease outbreak in North America. Berkeley recognized the work of Morren, Teschemacher, and other investigators in his 1845 paper, but was unaware of Abbé Edouard van den Hecke.

4. A. de Bary, *Journal of the Royal Agricultural Society of England*, 2nd ser., 12, 239–269, 8 figures (1876).

5. G. N. Agrios, *Plant Pathology*, 4th edition (San Diego, CA: Academic Press, 1997).

6. M. J. Carlisle, in *Water, Fungi and Plants*, edited by P. G. Ayres and L. Boddy (Cambridge, UK: Cambridge University Press, 1986), 105–118.

7. P. van West et al., *Molecular Plant-Microbe Interactions* 15, 790–798 (2002).

8. S. Kamoun and C. D. Smart, *Plant Disease* 89, 692–699 (2005). For comparison, 11,109 genes have been reported in the rice blast fungus, *Magnaporthe grisea*.

9. Effectors are encoded by a pathogen's virulence genes. The pathogen's avirulence genes (*Avr*) encode molecules that are recognized by the host (they are recognized by proteins produced by plant disease-resistance [R] genes), leading to the hypersensitive response. See B. M. Tyler, *Annual Review of Phytopathology* 40, 137–167 (2002), and P. R. J. Birch et al., *Trends in Microbiology* 14, 8–11 (2006).

10. T. A. Randall et al., *Molecular Plant-Microbe Interactions* 18, 229–243 (2005). In this analysis of the genome of *Phytophthora infestans*, genes encoding enzymes that the pathogen may use to dissolve the plant cell wall were similar to those found in Kingdom Fungi. The oomycete also encoded genes for the synthesis of chitin. This is interesting because chitin is often cited as a defining characteristic of the cell walls of species in Kingdom Fungi that is absent from oomycete water molds. This isn't a surefire distinction, however,

because small quantities of chitin have been detected in the walls of some oomycetes. The presence of "fungal-looking" genes for attacking the cell walls of plants, and for constructing the cells walls of the pathogen, may be explained by a process of convergent evolution. Alternatively, genes may have moved between these groups of microbes by horizontal transfer. If this explains the genetic matches between fungi and oomycete water molds, it is interesting to speculate on the direction of the transfer. Genes that encode the enzymes for the digestion of pectin within plant cell walls, for example, may have moved from ancient fungi to ancient water molds or vice versa.

11. J. O. Anderson and A. J. Roger, *Current Biology* 12, 115–119 (2002).

12. N. P. Money, C. M. Davis, and J. P. Ravishankar, *Fungal Genetics and Biology* 41, 872–876 (2004).

13. S. Kamoun, *Eukaryotic Cell* 2, 191–199 (2003); M. Latijnhouwers, P. J. G. M. de Wit, and F. Govers, *Trends in Microbiology* 11, 462–469 (2003).

14. American, Canadian, and European researchers share data through the Phytophthora Genome Consortium and Syngenta Phytophthora Consortium. Syngenta is a Swiss agribusiness company that is sequencing the genomes of many pathogenic fungi. Data from both organizations can be accessed from http://www.pfgd.org.

15. Buczacki (n. 2).

16. E. Large, *The Advance of the Fungi* (New York: Henry Holt & Company, 1940), 31. Large's reference to *Phytophthora* as a "weird and colorless seaweed" squares with today's understanding of the algal origin of the oomycete water molds.

17. B. Prévost, *Memoir on the Immediate Cause of Bunt or Smut of Wheat, and of Several Other Diseases of Plants, and on Preventives of Bunt,* trans. G. W. Keitt (Menasha, WI: American Phytopathological Society, 1939).

18. Berkeley identified the fungi collected by Charles Darwin during his voyage on the Beagle (1831–1836). His personal herbarium, donated to Kew in the 1870s, contained more than 10,000 species.

19. Like Tillet and Prévost's experiments, Pasteur's investigations were stimulated by a competition. In Pasteur's case, the French Academy of Sciences offered a prize to the scientist who could settle the issue of spontaneous generation. Pasteur claimed the prize in 1864.

20. Rather than assailing someone for a hearing disability, the insult is used against those who don't pay attention to imparted information.

21. Berkeley (n. 1), p. 24.

22. N. E. Stevens, *Journal of the Washington Academy of Sciences* 23, 435–446 (1933). Stevens' account of potato late blight covers the American epidemic of 1843–1845 and emphasizes the work of James Teschemacher, who was curator of botany for the Boston Natural History Society (see n. 3).

23. P. M. A. Bourke, *Nature* 203, 805–808 (1964).

24. C. Fogarty, *Irish People*, New York (October 26, November 2, 1996).

25. T. Woods and R. Kavana, "Young Ned of the Hill" (Stiff America/Happy Valley Music [BMI], 1989). In The Pogues version of the song, the broken-toothed Shane MacGowan sings, "A curse upon you Oliver Cromwell, You who raped our Motherland, I hope you're rotting down in hell, For the horrors that you sent."

26. L. Zuckerman, *The Potato: How the Humble Spud Rescued the Western World* (Boston: Faber and Faber, 1998).

27. De Bary (n. 4). In this paper, De Bary reviewed the results of his earlier work published in German in 1861 and 1863. The paper was commissioned by the Royal Agricultural Society, which persuaded De Bary to return to his work on potato blight when a new outbreak of the disease in 1872 threatened a recurrence of the famine in Ireland.

28. Charles Morren of Liége understood that the early removal of infected stems and leaves could save the tubers. Morren was an early supporter of the fungal theory of potato blight.

29. P. G. Ayres, *Mycologist* 18, 23–26 (2004).

30. P. M. A. Millardet, *The Discovery of the Bordeaux Mixture*, trans. by F. J. Schneiderhan (Ithaca, NY: American Phytopathological Society, 1933).

31. Mycologists recognize two types of mildew. Downy mildews are caused by oomycete water molds. Powdery mildews are caused by ascomycete fungi that form tiny spherical fruiting bodies called cleistothecia. Grapes are afflicted by both kinds of mildew. Downy mildew of grape is caused by *Plasmopara viticola*, and powdery mildew of grape is caused by the fungus *Uncinula necator*.

32. H. H. Whetzel, *An Outline of the History of Phytopathology* (Philadelphia: W. B. Saunders Company, 1918). Whetzel wrote that the Millardetian Period was marked by the recognition of the economic importance of the study of plant disease.

33. Inorganic fungicidal chemicals such as sulfur, copper, and mercury block the electron transport chain that generates energy in mitochondria. Copper also damages cell membranes and disrupts nucleic acids and proteins; G. Borkow and J. Gabbay, *Current Medicinal Chemistry* 12, 2163–2175 (2005).

34. F. J. Schwinn, in *Phytophthora: Its Biology, Taxonomy, Ecology, and Pathology*, edited by D. S. Erwin, S. Bartnicki-Garcia, and P. H. Tsao (St. Paul, MN: The American Phytopathological Society 1983), 327–334.

35. Bourke (n. 23).

36. N. P. Money, *Mr. Bloomfield's Orchard: The Mysterious World of Mushrooms, Molds, and Mycologists* (New York: Oxford University Press, 2002), 129–138.

37. W. G. Smith, *Nature* 12, 234 (1875). Worthington George Smith (1835–1917) was among the many victims of Cincinnati's eccentric mycologist Curtis Gates Lloyd (see Money, n. 36). Reviewing Smith's book, *Synopsis of the British Basidiomycetes: A Descriptive Catalogue of the Drawings and Specimens in the Department of Botany, British Museum* (London: Trustees of the British Museum, 1908), Lloyd wrote that it seemed "like an attempt by someone living in the Sahara to write a book about a rain forest."

38. De Bary (n. 4).

39. G. P. Clinton, *Report of the Connecticut Agricultural Experiment Station* 33–34, 753–754 (1911); G. P. Clinton, *Science* 33, 744–747 (1911).

40. G. H. Pethybridge, *Scientific Proceedings of the Royal Dublin Society*, n.s., 13, 529–565 (1913).

41. J. B. Ristaino, *Microbes and Infection* 4, 1369–1377 (2002).

42. S. B. Goodwin, B. A. Cohen, and W. E. Fry, *Proceedings of the National Academy of Sciences USA* 91, 11591–11595 (1994).

43. W. E. Fry and S. B. Goodwin, *Bioscience* 47, 363–371 (1997).

44. Q. Schiermeier, *Nature* 410, 1011 (2001). Russia is the world's second largest potato producer after China.

45. http://138.23.152.128/JudelsonHome.html.

46. J. B. Ristaino, *Phytopathology* 88, 1120–1130 (1998).

47. J. B. Ristaino, C. T. Groves, and G. R. Parra, *Nature* 411, 695–697 (2001); K. J. May and J. B. Ristaino, *Mycological Research* 108, 471–479 (2004).

48. The calculations are based on numbers in Carlisle (n. 6). Swimming zoospores consume about 1 picogram of proteins, lipids, and glycogen per hour.

49. Bourke (n. 23).

50. S. Heaney "At a Potato Digging," in *Poems 1965–1975* (New York: Farrar, Straus and Giroux, 1980). © Seamus Heaney, reprinted with permission.

Chapter 8

1. T. N. Taylor et al., *Transactions of the Royal Society of Edinburgh: Earth Sciences* 94, 457–473 (2004).

2. T. N. Taylor, H. Hass, and W. Remy, *Mycologia* 84, 901–910 (1992).

3. B. B. Kinloch, *Phytopathology* 93, 1044–1047 (2003).

4. R. P. Scheffer, *The Nature of Disease in Plants* (Cambridge, UK: Cambridge University Press, 1997).

5. W. V. Benedict, *History of White Pine Blister Rust Control—A Personal Account*, USDA Forest Service FS-355 (Washington, DC: U. S. Government Printing Office, 1981).

6. O. C. Maloy, *Annual Review of Phytopathology* 35, 87–109 (1997).

7. Thaxter wrote, that "Bordeaux mixture is the vilest compound imaginable, but it would give me some satisfaction to spray [some] Connecticut farmers with it until . . . the moss started from their backs" (p. 33). This quote is taken from J. G. Horsfall's biographical essay on Thaxter in *Annual Review of Phytopathology* 17, 29–35 (1979). Thaxter's employment as a plant pathologist was brief and preceded his academic career at Harvard, during which he catalogued an unusual group of fungi called the Laboulbeniales. His contributions to mycology are detailed by W. H. Weston, *Mycologia* 25, 69–89 (1933).

8. A similar disquisition about the internecine conflict between plant pathologists and mycologists appears in an 80-year-old book by plant pathologist Arnold Sharples: *Diseases and Pest of the Rubber Tree* (London: MacMillan and Company, 1936).

9. G. S. Gilbert, *Annual Review of Phytopathology* 40, 13–43 (2002).

10. This concept is also illustrated by another fungus, *Cronartium quercuum*, that causes fusiform rust of southern pines; R. A. Schmidt, *Phytopathology* 93, 1048–1051 (2003). The rust is maintained at low levels in old-growth stands of pine trees but has caused epidemic disease in plantations since the 1960s. Planting of rust-resistant pines, coupled with careful management of the trees through thinning and pruning, look promising in the fight against this disease.

11. J. Krakowski, S. N. Aitken, and Y. A. El-Kassaby, *Conservation Genetics* 4, 581–593 (2003).

12. D. P. Reinhart et al., *Western North American Naturalist* 61, 277–288 (2001).

13. J. Muir, *My First Summer in the Sierra* (Boston: Houghton Mifflin, 1911), p. 211.

14. Even older plants include creosote bushes in the Mojave desert, box huckleberry in Pennsylvania, and a eucalypt called ice age gum in Australia. Age estimates for these plants span 11,000 to 13,000 years.

15. J. T. Blodgett, *Plant Disease* 88, 311 (2004).

16. D. M. Rizzo et al., *Plant Disease* 86, 205–214 (2002). This was the first major work on SOD in which the authors marshaled an impressive body of research to unmask the cause of the epidemic.

17. S. Werres et al., *Mycological Research* 105, 1155–1165 (2001).

18. The British plant pathologist Clive Brasier, whose work on Dutch elm disease was discussed in chapter 2, made the initial connection between the American and European pathogen.

19. M. Garbelotto, P. Svihra, and D. M. Rizzo, *California Agriculture* 55, 9–19 (2001).

20. P. E. Maloney et al., *Plant Disease* 86, 1274 (2002); J. Knight, *Nature* 415, 251 (2002).

21. E. M. Goheen et al., *Phytopathology* 92, S30 (2002).

22. The California Oak Mortality Task Force web site (www.suddenoakdeath.org) is a superb resource for up-to-date information on the spread of SOD.

23. J. Withgott, *Science* 305, 1101 (2004).

24. P. Healey, *The New York Times* (July 29, 2004), p. A20. At the time of writing, researchers are carrying out additional tests to corroborate the presence of the pathogen.

25. Oak wilt is one of the many serious fungal diseases whose story is not detailed in this book. *Ceratocystis fagacearum* is related to the fungus that causes Dutch elm disease: both pathogens cause vascular disease and are spread by beetles. Although oak wilt kills thousands of trees every year, this represents a tiny fraction of the oak cover.

26. E. Stokstad, *Science* 203, 1959 (2004).

27. www.suddenoakdeath.org.

28. B. Henricot and C. Prior, *Mycologist* 18, 151–156 (2004).

29. C. Brazier, *Mycological Research* 107, 258–259 (2004).

30. C. L. Schardl and K. D. Craven, *Molecular Ecology* 12, 2861–2873 (2003).

31. C. Brasier and S. Kirk, *Mycological Research* 108, 823–827 (2004).

32. Henricot and Prior (n. 28).

33. G. Weste and G. C. Marks, *Annual Review of Phytopathology* 25, 207–229 (1987).

34. Less than 100 mature Wollemi pines, *Wollemia nobilis*, grow in a rainforest gorge 200 kilometers west of Sydney (www.wollemipine.com). *Phytophthora cinnamomi* has not been isolated from the soil associated with these trees.

35. *Threat Abatement Plan for Dieback Caused by the Root-Rot Fungus Phytophthora cinnamomi* (Canberra: Commonwealth of Australia, 2001), p. 12.

36. F. D. Podger, *Phytopathology* 62, 972–981 (1972).

37. www.calm.wa.gov.au/projects/dieback_phosphite.html.

38. Satellite mapping is discussed in the government *Threat Abatement Plan* (n. 35). The use of this technology in tracking SOD is detailed in M. Kelly, K. Tuxen, F. Kearns, *Photogrammetric Engineering and Remote Sensing* 70, 1001–1004 (2004).

39. Weste and Marks (n. 33).

40. www.apsnet.org/online/SOD/Papers/Brasier/default.htm.

41. S. Anagnostakis, *Biological Invasions* 3, 245–254 (2001). The disease also killed Allegheny and Ozark chinquapins, which are close relatives of the American chestnut.

42. B. S. Crandall, G. F. Gravatt, and M. M. Ryan, *Phytopathology* 35, 162–180 (1944); B. S. Crandall, *Plant Disease Reporter* 34, 194–196 (1950). When *Phytophthora cinnamomi* attacks chestnuts, it rots the root system, causing an inky-blue exudate. This symptom accounts for the common name "La maladie de l'encre," or "ink disease," in France, where the same plague destroyed orchards of European chestnuts in the 1870s. Ink disease continues to be a problem for commercial chestnut growers in Europe.

43. E. Stokstad, *Science* 306, 1672–1673 (2004).

44. Scheffer (n. 4).

45. R. S. Ziegler, S. A. Leong, and P. S. Teeng, *Rice Blast Disease* (Wallingford, UK: CAB International, 1994).

46. R. A. Dean et al., *Nature* 434, 980–986 (2005).

47. S. M. Whitby, *Biological Warfare Against Crops* (New York: Palgrave, 2002); L. V. Madden and M. Wheelis, *Annual Review of Phytopathology* 41, 155–176 (2003).

48. Brown spot of rice was partly responsible for famine in India in the 1940s, during which more than two million people starved.

49. P. Rogers, S. Whitby, and M. Dando, *Scientific American* 280, 70–75 (1999).

50. R. C. Mikesh, *Japan's World War II Balloon Bomb Attacks on North America* (Washington DC: Smithsonian Institution Press, 1973).

51. Detailed information on Iraq's program of chemical and biological warfare has been compiled by the Center for Nonproliferation Studies (http://www.nti.org) and by the Stockholm International Peace Research Institute (http://editors.sipri.se/pubs/Factsheet/unscom.html).

52. http://www.slate.msn.com (accessed October 3, 2002).

53. There is evidence of Iraqi research on other fungal toxins including trichothecenes that have been associated with toxic indoor molds; N. P. Money, *Carpet Monsters and Killer Spores: A Natural History of Toxic Mold* (New York: Oxford University Press, 2004).

54. V. Vajda and S. McLoughlin, *Science* 303, 1489 (2004).

55. A similar pattern of plant die-off, fungal spike, and plant reclamation is recognized in the fossil record from the Permian-Triassic boundary; M. J. Benton and R. J. Twitchett, *Trends in Ecology and Evolution* 18, 358–365 (2003).

56. A. Casadevall, *Fungal Genetics and Biology* 42, 98–106 (2005).

57. Mushroom-forming species can colonize the tissues of patients with impaired immune defenses. A recent case history from Texas described a young man infected by a species of *Phellinus* that is otherwise known as a wood-rotting fungus. D. A. Sutton et al., *Journal of Clinical Microbiology* 43, 982–987 (2005).

58. Although nobody has discovered the fossil remains of an inhaler between the lips of a tyrannosaur, perhaps the dinosaurs were offed by an allergic response to the blanket of mold spores.

59. Nixon, K. C. et al., *Annals of the Missouri Botanical Garden* 81, 484–533 (1994).

60. This idea can be credited to my colleague Roger Meicenheimer (unless he's right, and then I'd like the credit).

Index

Printed in the USA/Agawam, MA
October 22, 2012